The Evolution of Intelligent Systems

Also by Ken Richardson

COGNITIVE DEVELOPMENT TO ADOLESCENCE (*co-edited*)
UNDERSTANDING INTELLIGENCE
UNDERSTANDING PSYCHOLOGY
THE ORIGINS OF HUMAN POTENTIAL
MODELS OF COGNITIVE DEVELOPMENT
THE MAKING OF INTELLIGENCE

The Evolution of Intelligent Systems
How Molecules Became Minds

Ken Richardson

© Ken Richardson 2010

All rights reserved. No reproduction, copy or transmission of this publication may be made without written permission.

No portion of this publication may be reproduced, copied or transmitted save with written permission or in accordance with the provisions of the Copyright, Designs and Patents Act 1988, or under the terms of any licence permitting limited copying issued by the Copyright Licensing Agency, Saffron House, 6-10 Kirby Street, London EC1N 8TS.

Any person who does any unauthorized act in relation to this publication may be liable to criminal prosecution and civil claims for damages.

The author has asserted his right to be identified as the author of this work in accordance with the Copyright, Designs and Patents Act 1988.

First published 2010 by
PALGRAVE MACMILLAN

Palgrave Macmillan in the UK is an imprint of Macmillan Publishers Limited, registered in England, company number 785998, of Houndmills, Basingstoke, Hampshire RG21 6XS.

Palgrave Macmillan in the US is a division of St Martin's Press LLC, 175 Fifth Avenue, New York, NY 10010.

Palgrave Macmillan is the global academic imprint of the above companies and has companies and representatives throughout the world.

Palgrave® and Macmillan® are registered trademarks in the United States, the United Kingdom, Europe and other countries.

ISBN: 978–0–230–25249–3 hardback

This book is printed on paper suitable for recycling and made from fully managed and sustained forest sources. Logging, pulping and manufacturing processes are expected to conform to the environmental regulations of the country of origin.

A catalogue record for this book is available from the British Library.

Library of Congress Cataloging-in-Publication Data

Richardson, Ken.
   The evolution of intelligent systems : how molecules became minds / Ken Richardson.
     p. cm.
     ISBN 978–0–230–25249–3 (hardback)
     1. Brain – Evolution. 2. Molecular evolution. 3. Evolutionary psychology. I. Title.

QP376.R476 2010
612.8′2—dc22
                                                                      2010027546

10 9 8 7 6 5 4 3 2 1
19 18 17 16 15 14 13 12 11 10

Printed and bound in Great Britain by
CPI Antony Rowe, Chippenham and Eastbourne

Contents

List of Illustrations   vi

Preface   viii

1. Why So Complex?   1
2. Fit for What?   19
3. Intelligent Life   36
4. Bodily Intelligence   56
5. Evolution of Development   69
6. Intelligent Eye and Brain   90
7. From Neurons to Cognition   112
8. Cognitive Functions   135
9. Social Intelligence   160
10. Intelligent Humans   174

Notes   199

Index   229

Illustrations

Table

8.1 The XOR problem — 156

Figures

1.1 Heads of some of the finches studied by Charles Darwin showing the correlation between beak structure and environmental structure (type of food): (1) a large seedeater; (2) a small seedeater; (3) small insect eater; (4) large insect eater. (Redrawn from Darwin) — 9
1.2 Maturana and Varela's notion of structural coupling between organism and environment. (Redrawn from Maturana and Varela) — 16
2.1 Graph of interactive and non-linear relationships. The non-linear associations between variables A and B is conditioned by (varies with) values of variable C (C1-C3) — 24
2.2 Phase diagrams showing movements of a frictionless pendulum in a limit-cycle attractor (left) and a normal pendulum settling into a steady-state (or point) attractor (right) — 29
2.3 The Lorenz attractor is a graphic description of patterns of behaviour of complex systems like the weather — 30
2.4 Bénard cells forming in a layer of liquid — 33
3.1 Schematic diagram of *E. coli* to show receptors and flagella — 48
3.2 Simplified diagram of regulations in circadian rhythm of the fruitfly — 53
5.1 Getting connected. Axon terminals, guided by molecular signals, find target cells. There, development of dendrite terminals is promoted by structured spike firing along the axons — 88
6.1 A sequence of stills from a 'point light walker'. Volunteers find the images unrecognizable when presented individually, on a computer screen. But presentation as a sequence at normal speed evokes almost immediate recognition of a person walking — 95

6.2	Schematic diagram of retina and cell connections	97
6.3	Kanizsa triangle	98
7.1	Fragment of an artificial neural network. Stimuli (usually some well-defined features, as conceived by the experimenter) are input at the keyboard and received by the input units. If the strength of signal exceeds a certain threshold, the unit discharges along its connections to other units. The connections can be excitatory or inhibitory (e.g. a+/a−). After integration of a series of signals in subsequent units, the outputs should reflect statistical associations between the inputs	116
7.2	Extraction of covariation structure in early processing	121
7.3	Experiencing a 'line'	121
7.4	A new stimulus (arrows), moving as a wave of activation containing 'sample' feature parameters, is drawn towards basins of attraction containing corresponding sets of parameters ('contour' lines, or isoclines, connect parameters that have occurred with similar frequencies in previous experience, those at deeper levels representing the more coherent sets, e.g. the more 'definite' line image)	123
7.5	The Necker cube	127
8.1	Conservation of number over changes in appearance	149

Preface

Many people interested in the nature and origins of the human mind complain about the lack of coherence and depth in existing scientific accounts. They say that the area lacks a meaningful framework through which they can relate the various models, theories and opinions being expressed within it; and it lacks 'roots' in the way that evolutionary theory gives 'history', coherence and grounding to all of biology. It is a truism spoken by many that if we want to truly understand functions of living things – including mental systems – then we really need to be clear about why they evolved. Sadly, recent attempts to incorporate psychology within the classical evolutionary framework, far from yielding fresh insights, seem to have generated even more division and controversy. In consequence, there are still crucial gaps in our understanding of human minds and human nature, and about their continuity with the rest of life. This creates many problems for scientists, and increasing public scepticism about scientific accounts of what we've actually got in our heads and how it works.

The main reason for the gaps, I think, is the difficulty in envisaging and describing the conditions under which something as brilliant, and seemingly transcendental, as the human mind could have arisen from far simpler material beginnings. To many people it seems to have required something like a miracle, or the hand of an intelligent designer: what else could account for the amazing complexity of human minds and the gulf between it and all other living forms? Fortunately, that difficulty is now easing as new methodological and theoretical advances – especially in the realm of dynamic systems – begin to make a coherent picture of 'mind from matter' more feasible. It requires new concepts, from a wide range of disciplines, and new ways of thinking about the world, but the picture is transforming our views of science and evolution, as well as the nature and origins of mind. This book aims to bring those new concepts together, hopefully in a style accessible to most general readers, as well as serious students, and with widespread implications for our views of the origins and nature of mind.

This is, of course, a tall order, and a difficult trail to follow: a convincing story calls for more than mere description of a sequence of forms (a kind of 'tree of intelligence' to parallel Darwin's tree of life). It also requires an underlying 'plot', a causal rationale, or a description of the evolutionary 'laws of motion' wringing such a remarkable sequence from ordinary natural materials and forces. The realisation that the world is not the stable, orderly place that dogma once deemed it to be, but one of continual interaction and change – a very dynamic place – has also created new concepts of the complexity and structure of the world, and, indeed, the great beauty, that lies within. These ideas have brought new focus on environmental change and on the evolution of intelligent systems for dealing with it. So the story, here, traces intelligent systems from the origins of life in molecular ensembles, through succeeding levels of complexity in cells, brains and cognitive systems, to finally demystify the origins and nature of the human mind itself.

It has, of course, been a thrilling trail to follow and I hope readers too will share some of that excitement. It would not have been possible without the thoughts and inspiration of the many ideas-makers I have had the pleasure of encountering along the way, either in person, or by other means. They are too numerous to list, here, but I hope their presence in these pages and/or the nature of the product will reflect my gratitude. The following, however, took time out from their own busy schedules to read a draft of the present work and offer comment and feedback: Eric Matthews, Hellen Matthews and Paul Prescott. I have not been able to take all feedback on board, but most of it really helped. Dan Richardson did most of the diagrams, and my partner Susan gave me her usual unstinting support. I am grateful to all of them for helping me turn a rough draft into a more readable and coherent product. If it still isn't, the responsibility is, of course, entirely my own.

KEN RICHARDSON

1
Why So Complex?

An incomplete story

Charles Darwin's theory of evolution has answered so many questions about human origins over the last 150 years. Most people now accept that humans are the descendents of other species modified by natural selection from random variations. The trail is clearly there in shared structures, in the sequence of development in the embryos, in the fossil record and its changes over time, and so on. So, we can accept the story, more or less as Darwin told it, along with all its implications.[1] Yet modern biologists and psychologists know, as laypersons suspect, some aspects of the story are incomplete.

The problems start when we think about the more complex aspects of living things. Of course, everything in biology is recognised as complex. Many books have been written about the origins and nature of complex characteristics of species. Darwin himself proposed how physical features could become more complex over time by the work of natural selection, building through small gradations from simpler structures. We can trace the highly dexterous human hand from that of apes (with non-opposable thumb) and, before that, to homologous, simpler structures in feet, wings and fins. Darwin was able to propose the evolution of the complex eye of mammals from simpler forms, even back to light-sensitive cells in some primitive organisms.[2] Today, of course, we can trace some of our less visible, but even more complex, physiologies and metabolisms back to self-organising molecular systems in single cells. The continuity of succession in all these things helps us to understand why they evolved and, through that, their true nature.

What I'm particularly referring to, though, are the vastly more complex intelligent systems of animals, a kind of culmination of which we see in that most complex of all phenomena, the human mind. The complexities of sensing, perceiving, thinking, feeling, decision-making and cognitive systems seem to be on a different level altogether from physical, or even physiological, complexities. They presumably evolved from antecedent intelligent systems: but from what, and how and why, are still matters of great debate. We may be able to trace them back through the cognitive systems of apes and other mammals, as Darwin attempted to do in his *Descent of Man*.[3] But, beyond that, the trail becomes murky, and we seem to peer into a mist of uncertainty. With no clear signs of origins or continuities, we cling to Darwin's theory, almost as an act of faith, but with little exposition of what really evolved and why. And that, I believe, still leaves a void in both the scientific and public understanding of our cognitive system, in particular, and of human nature, in general.

The problem of continuity is, of course, one of tracing increasing complexity over long periods of time. So why is it so difficult to describe for intelligent systems? One reason, of course, is that intelligent systems are so much more complex to describe than the hand, or even the eye. Another reason is that the change in complexity must have been so much greater, so that antecedents are probably so different. And how far back should the thread take us? Before mammals? To single cells? To the origins of life itself? Moreover, we need more than mere descriptions: we also need *explanation* for their evolution, which usually means some sort of functional description of why they were favoured in the course of natural selection. These questions are not, of course, independent: not having that explanation for their evolution makes us uncertain about what to describe (even in contemporary systems), and vice versa.

As for explanation, then, some may think this is straightforward and pretty obvious, and rather like increase in any kind of complexity. It is favoured because it assists survival, as an intrinsic property of the character. It is an intuitively appealing idea that any part or function of a species will always tend to improve if it assists survival and there is natural selection. Indeed, Darwin suggested, reasonably enough, that complexity, with more numerous components and improved organisation, imparts greater 'efficiency' in the struggle for life.[4] So what's the problem, you might ask?

The problem is that the issue is not as simple as you might think, and there's an intense contemporary debate in biology about it.[5] Increase in complexity of *any* part or function isn't automatic because it usually incurs biological costs: so there have to be good net reasons for it. More complex parts or functions are more difficult to create and to develop and maintain, require more energy, and are more prone to catastrophic breakdown. In addition, apparent improvement in one part – say the size of a limb – affects the rest of the body, and may diminish the functioning of associated parts. It also usually requires the necessary genetic variations.

This cost explains why, for many organisms or characters, natural selection favours things being simple: a possibility also acknowledged by Darwin. After all, simple forms such as yeasts and bacteria still thrive, after billions of years, in their original habitats and, indeed, form the major proportion of the earth's biomass. We may ask, why haven't they, everywhere, been replaced by more complex forms? Indeed, over 85% of the 3.5 billion year history of life on earth was taken up with nothing but single-cell organisms. Then suddenly they started to become more complex; intelligent *systems* immediately started to evolve; and we aren't quite sure why or what from. Today, we know of many examples of evolution in which there is reduction, rather than increase, of complexity over time (parasites being the classic cases). So not all explanations for a progression are as clear-cut as those for the hand or the eye. We need convincing *reasons* for increased complexity, not just an available mechanism, a belief in the intrinsic superiority of complexity and faith that progress will automatically ensue.

Another approach to functional explanation, then, has been to look at the conditions – that is, the environment – that might have favoured increase in complexity. Unfortunately the search has been rather thin, and has been largely confined to 'arms race' stories. Such arms races, as between predators and prey, describe spirals of complexity between competing species: evolution of faster foxes leads to evolution of faster rabbits, demanding even faster foxes and so on. Will some sort of arms race explain the progression in complexity of intelligent systems – or of complexity in general? That kind of account has certainly been influential. For example, in *The Extended Phenotype*, Richard Dawkins says that 'it is arms races that have injected such "progressiveness" *as there is* in evolution' (my emphasis). And in *The Blind Watchmaker*: 'Progressive improvement of the kind suggested by the arms race image does go on, even if it goes on *spasmodically and interruptedly*' (my emphasis).[6]

The problem here is that arms races tend to revolve around rather specific characteristics (like running speed), and often, as with increased size, heavy shells, huge claws or teeth, at great expense to the general economy of the organism. Indeed, Thomas Miconi reviews a number of ways in which evolution of simplicity is better than increased complexity, even in arms races.[7] Arms races, that is, produce no *general* progress in 'accomplishment' (to use Dawkins' phrase): for example, the evolution of single cells into multicellular organisms; from asexual reproduction to sexual reproduction; from squashy worms and snails into vertebrates and functional integrations in physiology or the immune system. This is the kind of increased complexity that we seem to find emphatically in intelligent systems.

In consequence there is much uncertainty, even in biology generally, about why complexity like that in intelligent systems has increased. A blind faith in natural selection cannot be sustained because, 'natural selection theory by itself cannot account for increases in structural complexity'.[8] The lack of environmental or ecological analyses has produced claims that 'there is no attempt in neo-Darwinian theory to explain the ever-increasing complexity of living things'.[9] We need good reasons for increasing complexity – in intelligent systems, as all else – because, without them, 'the mechanisms behind the complexification and its relation to evolution are not well understood'.[10] If we aren't clear about such reasons with regard to complexity in general, it is hardly surprising that they haven't been clear with regard to intelligent systems.

This state of affairs has, of course, been disappointing for anyone – scientist or member of the public – trying to understand the origins and nature of highly complex systems like those that make up the human mind. But overcoming that disappointment is what this book is about. I think it is now feasible because of a host of new discoveries over the last two decades. These have revealed much more about the nature of environmental/ecological experiences, and provided better descriptions of complex systems themselves. They are indicating that the complexities are far more prominent, and the differences between them so much greater, than previously imagined. Indeed, discoveries of intelligent systems in previously unsuspected evolutionary levels have lead to demands for a revised or 'extended' evolutionary synthesis.[11] All this is making a clear picture of the evolution of complex intelligent systems from antecedents more feasible than ever: not just of a sequence of complexities, but of reasons for it. Before considering those reasons more closely, let us look at the urgency for doing so in both public and scientific spheres.

Public doubts

These issues are very important for the biological community, as we have seen. But my main worry is about broader consequences in the human sciences; in deeper questions in psychology, the nature of cognitive functions, of human development and the nature of humanity. Not having a clearer picture of the evolution of complex intelligent systems leaves psychology and cognitive sciences floundering without definite roots or adequate conceptions of current structures and functions. Although Darwin predicted that his theory would put psychology on a firmer footing, that has only partially happened. We still aren't sure how our intelligent, cognitive system is different from those of other living forms such as apes, nor the nature of the succession from reptiles, flies, or worms, or from other complex systems before that, even from non-living matter. Answers to date have produced, not a coherent image of a whole human mental system in evolution, but fragmented pictures that often seem to demean the beautiful complexity that we all sense is there.

What seems to make these matters particularly important is that the uncertainties surrounding them have affected the general public, not just scientists. They have raised doubts about the powers of science altogether. Apart from despair in actually describing what we've got, what it's for, and how it has originated, the gaps are sometimes seen as ones of scientific credibility in general, questioning even our ability to answer such questions. If science cannot describe how complex systems like our own cognitive abilities evolved, then perhaps we need to look elsewhere? One act of faith may be as good as another.

One expression of this uncertainty has been a new upsurge of creationism on both sides of the Atlantic: the belief that intelligent systems, especially human mental ones, can only be explained by the intervention of a supernatural 'intelligent designer' (though the issue of the infinite regress involved in that argument – i.e. about the origins of the designer – seems to be dodged). Recent support for that view has included proposals to re-introduce notions of 'intelligent design' into school curricula as alternatives to Darwin's theory of evolution.[12] So the old tensions between faith and authority, on the one hand, and science and reason, on the other, are reappearing. Even in this social sphere, then, a convincing story of the origins and nature of the human mind in all its complexity is long overdue.

Scientific doubts

The origins and nature of human mentality are ones of complexity *par excellence*, but the scientific answers have been neither complete nor convincing. Increasing numbers of scholars and scientists are realising and admitting that. Psychologists long ago turned to biology and Darwin's theory for insight. But, with the realisation that this remains the last great challenge in science, psychologists have, over the last twenty years, been joined by a host of others: evolutionary biologists, geneticists, molecular biochemists, embryologists, primatologists, behavioural scientists, neuroscientists, computer modellers, linguists, physicists, mathematicians and others. Their intense combined effort – much of it as determined and hard-nosed as anything in the history of science – has produced major methodological advances, a glorious profusion of findings, some remarkable insights, fascinating scientific models and more than a few bloated claims. But, there's still no generally agreed theory, no scientific conceptual framework to bring cohesion to the findings, and the big questions about our origins and true nature still attract much disagreement and controversy.

So the feeling that something important is still missing in both biology and psychology is being voiced in all fields. As Chris Sinha puts it, 'something happened...that radically transformed the evolving mind', and not knowing what it was 'poses a profound and complex problem for biological theory'.[13] Peter Richerson and Robert Boyd acknowledge that '[t]he complex cognition of humans is one of the great scientific puzzles.'[14] They note how an evolved cognitive system seems to explain our extraordinary success as a species. But they can only admit to 'our present state of ignorance' about why and how such complexity came about. In a later work they note the widespread view of our species as 'a spectacular evolutionary anomaly', and 'the evolutionary system behind it as anomalous as well'.[15]

This uncertainty is reflected even in theories about the basic nature of cognitive functions. Many ingenious models have been proposed for fundamental cognitive processes such as learning, memory, thinking, language and constructive action. But they notoriously fail to capture the true depths of human abilities: the depths and intricacies of our knowledge; the logical structures of everyday thought; the acoustic weave of every spoken utterance; the creativity of human cooperative endeavours; the bubbling ideas, imaginings and social playfulness of even a 5-year old – in fact, the whole nature of intelligence in animals and humans.

Let me offer just a few examples. Richard Shiffrin, a leading memory researcher, says: 'None of the models we use in psychology or cognitive science, at least for any behavioral tasks I find to be of any interest, are correct.'[16] Jerry Fodor, a leading cognitive theorist, says that failure to deal with mental functions in realistic contexts has put the field in 'deep denial', and that psychologists in truth still have very little idea of 'how the mind works.'[17] The doubts are repeated in the reflections of prominent psychologists recently presented in *The Psychologist*, the journal of the British Psychological Society (December, 2008): 'None of them felt there were adequate ideas or theories in place for understanding human nature...pointing out that the structural foundation of the science of psychology is incomplete, and in need of rebuilding'.[18]

The uncertainty prevails, it seems, on all fronts. It's usually assumed that intelligent systems must operate on the basis of knowledge, as opposed to fixed cues or signals (which is what distinguishes intelligence from instincts). The nature of knowledge, therefore, is supposed to be at the roots of cognitive theory. Yet, in his book, *How the Mind Works*, Steven Pinker mentions how psychologists feel 'perplexed' about the nature of knowledge, describing it as one of the problems that 'continue to baffle the modern mind.'[19] In his review of the subject, Emmanuel Pothos says: 'Overall, there has not been a single dominant proposal for understanding general knowledge.'[20] Many others complain about the fragmentary picture of the mind, with the kaleidoscopic vision of models, theories and 'approaches', we now have.

Moreover, in the intensity of scientific focus on ever-narrower models, the wider experiences of human existence in social pleasures, in emotional and cultural expressions, and appreciations of creative and natural beauties, have been sadly neglected. For example, a special series of articles in the journal *Nature* in 2008 notes how the appeal of music has defied all attempts at mathematical and scientific explanation.[21] So, from the perspective of such models, one of the most characteristic of human experiences seems arcane. Conversely, I hope to show that there is something like music nearly everywhere in nature.

I believe all of these expressions reflect the absence of that clear picture: understanding what kind of complex system we've got by describing how it emerged in evolution. Melanie Mitchell puts it neatly when she says that we lack 'well-worked out theories of evolutionary mechanisms that can explain psychological phenomena in a satisfying way. It might be said that the grandest open problem in evolutionary theory is to develop a unified theory of evolutionary mechanisms that could help settle these disputes.'[22]

This is partly the purpose of this book. Before I try to actually identify what is missing, though, let us have a look at attempts to use evolutionary theory in cognitive psychology so far, and see what has been wrong with them.

Evolutionary psychology – so far

The basic principle of natural selection is that the distinctive aspects of living things, including their more complex ones, have evolved as 'adaptations' to important parts of the environment. Much of that is evident from common observation: there are clear relationships between the characters (traits or characteristics) that living things have and particular aspects of the environments in which they live and have evolved. Beaks and mouths and teeth are formed according to the structure of the food their owners feed on. Outer skins are augmented with fur, feathers, or just left naked, according to the range of outside temperatures encountered. Limbs are shaped as legs, wings or flippers according to the medium or substrate in which they need to create movement.

As the theory goes, this kind of correspondence or 'fit' comes about because living things can accidentally sustain slight variation in their constitutions. These give rise to outward variations: either quantitative variation, such as size, or categorical variation, such as coat pattern, that can be inherited by offspring. One of these varieties may be (perhaps ever so slightly) more suited to an important aspect of the environment than others. They may be better protected against cold, for example, or against predators in their neighbourhood. So their bearers are more likely to survive and have more offspring. Through inheritance, the offspring will tend to have that same variety – and so on through successive generations. The net result of this differential survival is that the relative frequency of that variety increases in the population, and is recognized as characteristic, or a character, of a species. This is natural selection.

Evolutionists express the process in terms of selection pressures: that is, the particular demands of the environment which favour, and therefore shape, a particular variety to produce its final form. The classic cases of adapted structure are the finches studied by Darwin himself, with beaks shaped by natural selection for a particular kind of food (Figure 1.1). Evolutionary studies have become full of examples of well-adapted physical structures and physiological functions. Of course, we now know how such variations can appear to arise from variable genes, the biochemical structure of these – that is, DNA – also showing how this is possible. It is the genes that pass from parents to offspring.

The idea has been tremendously successful in accounting for most of the important physical features we have: so why not our mental ones, too? For over a 100 years, indeed, there have been numerous attempts to do just that, and these have intensified over the last couple

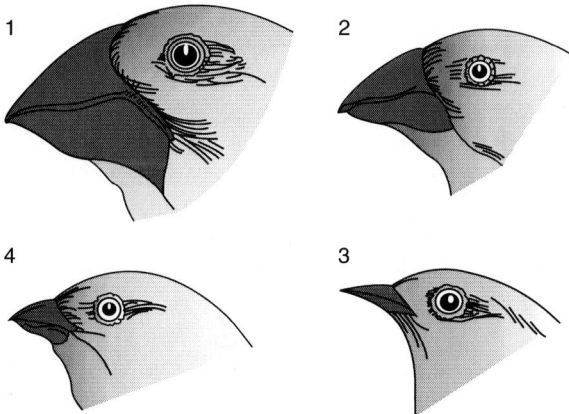

Figure 1.1 Heads of some of the finches studied by Charles Darwin showing the correlation between beak structure and environmental structure (type of food): (1) a large seedeater; (2) a small seedeater; (3) small insect eater; (4) large insect eater. (Redrawn from Darwin)
Source: Darwin, C. (1839) *Journal of Researches*, London: Colbourn.

of decades. By assuming that cognitive and other psychological functions can be treated as analogues of finch's beaks, investigators have tried to explain the origins and nature of mental functions through a Darwinian framework. This has established a new sub-discipline duly called evolutionary psychology (EP), extolling strong Darwinian principles. For example, Leah Cosmides and John Tooby have argued that, 'Natural selection shapes (mental) mechanisms so that their structure meshes with the evolutionarily-stable features of their particular problem-domains.'[23] On similar grounds, Steven Pinker argues, 'The mind is organized into modules or mental organs, each with a specialized design that makes it an expert in one arena of interaction with the world. The module's basic logic is specified by our genetic program. Their operation was shaped by natural selection to solve the problems of the hunting and gathering life led by our ancestors in most of our evolutionary history.'[24]

Such EP accounts, often tinged with a sensationalist element, have become very popular, even among the media and general public. So we read about quirks and vices of human behaviour such as crime, cheating, greed and gender differences, as only doing what comes 'naturally'. In psychology the appeal is now huge: a recent review, for example, has over forty chapters.[25] Increasingly, cognitive and other psychological traits have become 'explained' by analogy with biological traits shaped by natural selection.

But there are fundamental problems with such accounts. The main problem is that the account 'explains' adaptation to stable or durable environments, as the simplest reading of the theory demands: but stable environments tend not to be very complex. They are not very complex, because environments of Darwinian selection are necessarily rather constant, repetitive, ones. As you can probably see from the account so far, for natural selection to work, the aspect of the environment in question should persist across generations, or at least change only very slowly: without that continuity there could be no consistent selection. This is why Cosmides and Tooby refer to 'stable features' of the environment, and Pinker too implies the same. The assumption of little change and slow selection follows from Darwin. In chapter four of *The Origin of Species*, he emphasised that 'natural selection generally acts with extreme slowness (depending) on physical changes, which generally take place very slowly' and 'only at long intervals of time'.

So the evolutionary psychology programme has meant attempting to define the origins and nature of 'complex' cognitive functions in terms of relatively 'simple' environments. It is reflected in the notion, in EP accounts, that the mind consists of fixed mental 'organs', or computational devices, usually called modules. According to EP theorists, these have become functionally honed for specific, repetitive problems posed by specific aspects of the environment. One implication is that fixed modules will always tend to process information in the same way, according to the same set of rules (at least at some level). Another is that, since these modules were selected as adaptations to conditions prevailing in our hunter-gatherer period of evolution, they may be rather poorly adapted to any other, including our current, conditions. Thus it has been suggested that our skulls still carry essentially stone-age minds! It is such restrictive ideas that have drawn so much ire to evolutionary psychology from other scholars.

Environmental change and complexity

To be sure, the EP assumption about constant, or only slowly changing, environments across generations is reasonable enough with regard to some characters. The shapes and sizes of seeds are fairly constant and, therefore, predictable targets of selection for beak shape across generations: a variety that confers advantage now can be safely passed on to offspring (the form is predictable). Likewise with adaptations to a myriad other objects and forces encountered in the environment.

However, closer consideration of the real world indicates that, in spite of many classic cases, there are really not that many constant, repetitive environments at all. We have been increasingly realising that stable aspects of worlds, required by the simple evolutionary logic, are actually quite rare. Most aspects of environments are constantly and rapidly changing, often by the activities of organisms themselves. Sometimes radical changes can take place within the lifetimes of most organisms. We now know that the world of durable environments envisaged by Pinker, Cosmides and others, although true for some traits, are not typical. Changing environments produce a complexity of experience over time and space in a different league from Darwinian stable conditions. And it is the main purpose of this book to explain how the evolution of complexity in intelligent systems is related to that (hitherto under-described) complexity in the real world.

As we shall see, the relevance of this point hinges on the fact that changing environments contain structures or patterns that can be very useful to living things.[26] It seems strange to me that so many scientists have overlooked such complexity in environments. Even psychologists have tended to minimise the existence of such structure in experience, in spite of its relevance in understanding the evolution of complex functions. For example, Andy Clark and Chris Thornton admire the way that humans have a 'baffling facility' for problems in which the predictability lies in 'deeply buried regularities'. Yet they also claim that such problems are relatively rare.[27]

My argument here is that deeply buried regularities in changing environments describe the norm of everyday experience for most species. It is such aspects of some environments that make them more complex and lead to the evolution of more complex, intelligent systems. Such environments demand a kind of character different from stable fixtures and functions, or automatic responses. They need to be systems that allow the abstraction of deeper regularities, in some cases on an *ad hoc* and continuing basis, throughout life. Rather than relying on 'blind' natural selection to have shaped adaptations, the organism must be able to abstract information and generate predictions, and novel responses, on its own: that is, be *adaptable*, not merely adapted. So, it is complex, changeable, environments we need to focus on if we want to explain the evolution of complex, intelligent systems.

Describing environments

It seems likely, then, that complex functions evolved because they were demanded by complex environments. I start to look more closely at what that means in Chapter 2. For the moment it seems important to show how such environmental description has been either neglected or inadequate. The tendency, as with most EPs, has been to simply assume a role for more complex environments, with little further specification or demonstration of how they are actually connected with complex traits. The problem with describing complex environments within the simplest Darwinian model is, of course, the tendency to do the opposite, and simplify them.

We can illustrate this tendency to simplify the complex by looking at the model proposed by Peter Godfrey-Smith. He says, reasonably enough, that more complex forms of life and their cognitive systems emerged when they offered competitive advantage in more complex environments.[28] According to Godfrey-Smith, though, a complex environment is one that merely shows 'heterogeneity', or the 'range of possible states' that can be experienced over time and space. This environmental heterogeneity, he argues, can only be survived by greater heterogeneity (i.e., complexity) in the organisms adapting to it. So 'An organism is complex to the extent that it is heterogeneous.'[29]

However, the complexity – in environments and in organisms – being proposed, in that way, does not seem to be all that great. According to Godfrey-Smith, and many others, a complex environment is one that is experienced as a range of deviations around some ideal or 'normal' state. Departure from the ideal is signaled as a 'perturbation' on the organism to which it must react to regain its equilibrium or homeostasis. A complex organism, then, is one that has a lot of such equilibrium-regaining functions corresponding with different (changeable) aspects of the environment.

Such an account, of course, reflects a limited view of complexity of both organisms and environments. The classic analogy is that of fluctuating temperature on the outside corresponding with a heating mechanism on the inside turned on and off by a thermostat. It has been very difficult for theorists to explain how our minds have evolved from such simple environments. On the other hand, imagining the environments that *could* produce complex cognitive and brain functions has not been easy. Peter Richerson and Robert Boyd, for example, had to admit that, 'it is not entirely clear what selective regimes favor complex cognition.'[30]

Others have warned that, 'Most of the rules governing the evolutionary process toward more complex brains are still unknown.'[31] And a recent review concludes that, 'Human cognitive abilities are extraordinary[.] ... The conditions favoring the evolution of human cognitive adaptations remain an enigma ... it has proven difficult to identify a set of selective pressures that would have been sufficiently unique to the hominin lineage.'[32]

Part of that difficulty may lie in people's tendency to view natural selection as the work of a quasi-purposive agent making positive choices according to a pre-ordained scheme. This is the reason that some evolutionists are suspicious of notions of 'progress', such as increasing complexity, in evolution. But natural selection is not a process of making positive choices. It's not like shopping in a supermarket where you select goods from a shelf to meet preconceived criteria; rather it's a negative elimination of what you definitely don't want and having what's left. ('Natural rejection', or the 'filter' analogy, may, in fact, be better descriptions). It stands to reason that, in such a process, what will be rejected will be the odd and the misfit, as well as the downright damaged goods. What will be left will tend to be those things that are integrated with, or even improve, the harmony of an existing scheme.

Accordingly, the hand is selected, not as an independent unit, but against the background of a well-structured musculo-skeletal system within which it must functionally blend. Likewise, the selection of the eye needed to be integrated within nervous and muscular systems that could actually make it useful. (Darwin, in fact, noted how selection tends to produce such 'correlated features'). This of course puts a premium on any process within the organism that can help generate well-integrated novelties, rather than just random mutations. But, as we shall see, the process is helped by the fact that the environment itself is not experienced as a medley of independent goods on a market shelf. Rather it tends to be structured and patterned in predictable ways: even the experience of light is structured, as we shall see in Chapter 6. And that structure provides information for organisms for forms of engagement more intelligent than random mutation and passive elimination. So what evolves is a harmonious and dynamic marriage of structured system with structured environment.

One important task, then, seems clear: what we still seem to need is a more detailed view of the environmental conditions under which complex functions evolved – from origins of life to cognitive functions. Others have identified that task. Susan Oyama has mentioned how we need to 'consider again the endlessly fraught boundary between inside and outside and the attention given to the ill-defined external world.'[33] Massimo Pigliucci says that 'we are at a loss' when we try to 'pinpoint the biologically relevant components of an organism's milieu'.[34] In the context of human development, as Cathy Dent-Read and Patricia Zukow-Goldring say, 'defining what we mean by environment and "finding it"...appears more demanding.'[35] It's as if, in our haste to describe the inner workings of living systems we have tended to downplay the contexts in which they function. As Alan Love put it, we have encouraged 'the simplification of environmental causal factors in favor of isolating causal import from intrasystemic components'.[36]

As Oyama also says, it is not enough to just invoke interactions with the environment to understand complex structures and processes. We need to be more specific about the detailed nature of the environments that have resulted in more or less complex systems. Oyama quotes Ahouse and Berwick, who point out that, 'over 99 percent of our evolutionary history was spent in (and most of our genes arose in) a warm, salty sea.'[37] We clearly need to know what else it was about sea environments that produced such a range of complexity of beings, including some possessing advanced cognitive systems. Such inquiry may also tell us how and why cognitive systems are different from a number of other evolved complex systems. And it may help us realise why it is not enough to merely allude to 'plasticity' of structures and functions, or imply that the mind arrives in the world as a *tabula rasa* or 'blank slate'.

From reduction and determinism to dynamics

Of course, we can excuse the deficiencies so far by arguing that this is a very complex problem, so it takes a little longer to solve, even for the kind of mass assault we have been witnessing lately. But it is also possible that an entirely new *approach* or framework is needed. The problem is perfectly well illustrated in cognitive science where we now have mountains of specific 'findings' about very specific cognitive processes with little integration between them. This fragmentation has reflected the 'reductionist' tendency in all sciences. Reductionists, perfectly reasonably, break down and analyse a system in terms of its component parts and use the aggregate to explain the workings of the whole, in a mechanical sense – just as we might seek to describe the function of a strange machine by working out what the components do.

This strategy, in cognitive science, explains the increasing turn to sciences more basic than psychology itself, including brain sciences, evolutionary biology, genetics and so on.

The problem with reductionism is its tendency to lose sight of complexity. As Alan Love points out, the reductionist bias has led to narrow problem agendas such that higher levels of function have been neglected for the sake of identifying elements, or 'basic' processes.[38] Accordingly, many bright ideas stand out – but stand alone to give little general enlightenment. Perhaps the point has been put most succinctly by Gerald Edelman when, with reference to all the work on higher brain functions, he complains that 'relatively little progress has been made to integrate the results of this work into a global synthetic view of how the brain works.'[39] So it's perhaps not surprising that the reductionist approach to complex systems leaves major gaps in accounts of their evolution.

Of course, there have been attempts to produce more 'holistic' models of complex functions. But these, too, tend to have assumed fairly simple environments and mechanistic, homeostatic analogies. One popular example is the 'autopoietic' (self-producing) theory of Humberto Maturana and colleagues.[40] In a generally much admired thesis Maturana and Varela envisage an organism in constant interchange with its environment. However, they see the relationship as one of 'mutual congruent structural changes' resulting from 'reciprocal perturbations.'[41] They use a simple diagram to illustrate their notion of the relationship (Figure 1.2), where the wavy line implies a kind of periodically changing environment with which a living entity interacts. Such a conception of the environment, though, understates the complexity of both environments and intelligent systems. The account needs to explain why some organisms have become so much more complex than others.

The same tendency to simplify environmental complexity (and with it, the complexity of living systems) is seen in James Lovelock's famous model, called 'Gaia'. In the model, plants and animals react to environmental variables and the environment reacts to changes created by living things. So the two sides mutually correct one another's perturbations in a grand self-regulating system ('Gaia' is a term the Ancient Greeks used for their concept of Earth Goddess).[42] As a result, the whole ecosystem behaves as if a giant, self-organising, super-organism, just as the internal processes and interactions in living bodies maintain a constant temperature, blood pH, electrochemical balance and so on. Overall homeostasis is only disrupted by the development of extreme

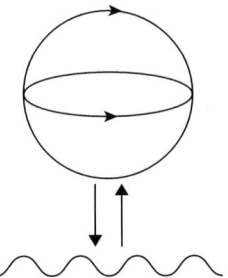

Figure 1.2 Maturana and Varela's notion of structural coupling between organism and environment. (Redrawn from Maturana and Varela)

values in one or other variable (as could now be happening through the excessive activities of humans). Lovelock's disappointing conception of the nature of the environment, however, is seen in his reference to 'the dull old real world of Nature which is most of the time either close to equilibrium or at a dynamic steady state.'[43]

The problem has been compounded by the fact that reductionism has gone hand in hand with a 'deterministic' scientific tradition. Deterministic models rely on fixed mechanisms that are, like machines, optimized to produce very predictable outcomes from narrow sets of known inputs or circumstances. Nearly all contemporary models of cognitive systems have been like that. It has been pointed out many times how deterministic, mechanical models can adequately perform simple tasks, but have difficulty with tasks of higher complexity.[44] Yet no one can deny that such higher complexity has spontaneously evolved numerous times since life began.

From stable to dynamic complexity

So the current book is an attempt to overcome the limitations in accounts of the evolution of complex systems, and their culmination in the human cognitive system. I believe that a more plausible picture of the emergence of increasingly complex intelligent systems in evolution will bring us a better understanding of their true nature. Such an attempt has become more possible because of a recent outpouring of new ideas about complexity and complex environments. Most of these now constitute a 'dynamic systems' approach to both complex environments and complex living systems. Much of this book will involve such models, and trying to make them generally comprehensible.

The application of such a view to the origins and nature of cognitive systems, I believe, helps fill-in the conceptual gaps that have created so many uncertainties about the origins and nature of mind and of humanity. Spinning out that argument, then, from complex dynamic environments to complex systems, from origins of life to human cognitive systems, is what this book is about.

Fortunately, there is now much to go on, and the reductionist, fragmented-mind, phase is changing in many branches of science. The invasion by a multitude of experts from other disciplines was bound to create a great deal of cross-fertilisation of ideas. This, in turn, is creating more general levels of concepts, including a better appreciation of the dynamic, or changeable, conditions of life. This has already become evident in biology. As evolutionary biologist David Depew says, we are entering a period in which there is 'renewed stress on the active, inventive, constructive, anticipatory, or behaviorally plastic nature of organisms' and on the role of these traits in *guiding* evolution (not just being blindly produced by it).[45]

Already, then, we have the beginning of a new tide of 'holistic' theories – for example, 'evo-devo' (evolution as influencing by development), or 'eco-evo-devo' (the need to consider more closely the details of the environment of a species).[46] The search for integrative models has drawn strongly on the study of complexity in several contributory disciplines, including computer modelling, mathematics, genetics, development and so on, and wonderful things have come out of it. These offer a more comprehensive understanding of evolution from ordinary non-living matter, through the origins of life to the emergence of complex cognitive systems.

Much of the excitement has stemmed from a closer look at the nature of experience in the real world, revealing just how much dynamic structure is there to foster the evolution of complex systems. The new field of dynamic systems theory (DST), sometimes under other guises such as non-linear dynamics, or the dynamical approach, is also showing that, in realistically changeable environments, with which most systems in living things have to cope, we need to focus on structures, not elements, in experience, in order to understand what has evolved. This has brought exciting new outlooks on living systems generally. In this book, I hope to show how they can portray evolution as a series of bridges or cascades, each responding to the dynamics of complexity in the world, and amplifying potential for complexity in living things. The trend culminates in cognitive powers that transcend the old Darwinian laws, in adapting the world to themselves, rather than vice versa. Little wonder that Eva Jablonka and Marion Lamb say that the first decades of the twenty-first century will be seen as revolutionary years for evolutionary theory, and that 'a new transformed Darwinian theory is upon us. It is time that the public knew more of this excitement and its significance'.[47]

2
Fit for What?

The previous chapter argued that, if we really want to understand the evolution of complex systems, then we need to look more closely at environmental complexity. What we see might help us understand, not only how increasingly complex systems have evolved, but also the *nature* of those systems. The purpose of this chapter is to explain how we have tried to understand complexity. First, I will consider some traditional conceptions of complexity, and of where it comes from. Next, I will consider briefly how mathematicians and scientists have described complexity in a more formal sense. Then I will consider how this has led to a revolution in conceptions of complexity. Finally, I will describe how these concepts are revolutionising so many aspects of science, our views of the world and of the nature of living things in it.

In her book, *Complexity*, Melanie Mitchell describes how scientific study in the area involves, 'explaining how large numbers of relatively simple entities organise themselves, without the benefit of any central controller, into a collective whole that creates patterns, uses information, and, in some cases, evolves and learns'.[1] Elsewhere, she points out that a suitable definition must also include other concepts, and later I will show how complexity is tied up with related notions such as structure and information.[2]

I will, in fact, argue that the essence of complexity is the 'organisation' that results from being 'structured'; that the structure has 'information' valuable to organisms and that, by evolving to assimilate that information organisms, too, become complex. I hope all this become clear in due course. In any case we all have some awareness of complexity in the natural world. In some cases it may be quite static, as in a mountain range, a salt crystal or large rock. Other complex forms are patently more dynamic (moving, changing), such as the cycles of our solar system, seasonal and daily rhythms, waterfalls, and clouds, and appear more organised. Although their specific states vary over time, we usually intuit some governing constraints, or 'rules', underneath the changes that are, themselves, relatively permanent and make up the 'system'. We may be more vaguely aware that more complex order, obeying more complex rules, exists in natural phenomena such as storms and hurricanes, chemical reactions, ocean tides and currents. Most of the time we don't think about how it originates: we just assume that it's part of a natural order.

Complexity by design?

When people *have* thought about where natural structure and its 'rules' come from, a curious paradox appears: this has been the tendency to try to 'explain' complexity as the product of some other, more complex, or even 'intelligent' agent. This tendency has probably been around since people first started to think about it. In Ancient Greece, Heraclitus proposed the existence of an 'ordering force', or energy, in non-living structures: a 'logos' that converts the random motions of the elements (earth, fire, water and air) into beauty and harmony. Around 340 BC Plato attributed the cosmic order to divinely invented laws of geometry to which all else, including human minds, conforms. His former pupil Aristotle, in observing cyclical events in nature introduced the influential notion of balance or 'equilibrium', so that nature itself was imagined to be a rational organism obeying necessary and purposeful rules.[3]

The tendency continued through the middle ages into modern times, and even into modern science. From the scientific revolution of the seventeenth century, it has been assumed that structural laws were somehow installed in the universe at its origins, and many have believed these were designed by some kind of superior intelligence.[4] The task of science was to describe the laws not their origins. Corresponding with this view has been a tendency among biologists and psychologists to think of 'structure-makers' in living systems, creating all their adaptations, from anatomical structures to complex physiological and cognitive functions.

Most recent of these structure-making 'ghosts in the machine', of course, has been the modern idea of the gene. Susan Oyama has drawn attention to the quasi-supernatural powers often accorded to genes, particularly by psychologists.[5] Alex Mauron explains how a 'scriptural' metaphor of an internal controlling agent has been imposed on the genes.[6] The publicity surrounding the sequencing of the human genome is further shaping contemporary ideas about how our genes allegedly make our humanity. Perhaps, in the absence of a wider sense of security in their existence people like the idea of some sort of program inside, with *something* in charge. But recent findings, as we shall see, are challenging that tendency.

Awareness of complexity and structure

Environmental complexity is an intrinsic and ubiquitous aspect of our experience. But it is formidably more difficult to define and describe than surface elements. Unfortunately, the appreciation of complexity among scholars, attempting to explain its role in evolution has tended not to be very deep. We can all enjoy the music we listen to as a manifestation of deeper structure. But describing its structure objectively or scientifically is a different matter. Such difficulty – as well as the reductionist tendency for simplicity – has inclined us to minimise the very existence of structure in experience.

As we saw in Chapter 1, one idea of complexity in nature is mere variability or heterogeneity. That idea describes complexity in living systems as correspondingly varied If-Then, or cue-response, functions such as the thermostat. But a number of other ideas have been produced. Richard Potts, for example, argues that the complexity of highly evolved species lies in their ability to adjust constantly to an *erratically* changing environment.[7] Similarly, Peter Richerson and his colleagues suggest that erratic changes in temperature, rainfall, atmospheric gases and their effects on vegetation, and so on, required greater behavioural flexibility in our ancestors.[8] Such accounts neither explain what, exactly, those abilities and flexibilities are, nor how they are actually related to the environment: there is little allusion to 'organisation' or 'structure' in them.

This seems surprising because some idea of structure has been around among philosophers, psychologists and other scholars for centuries. Again, philosophers remind us how Plato in Ancient Greece (and others) railed against elementism and saw the natural structures in nature as 'divine harmonies'. Plato was also aware of some similarity between music and mathematics, the 'language' of structure. Aristotle marvelled at the structure and organisation in living things, especially as they emerge and form in the embryo. Immanuel Kant, in the eighteenth century, insisted that the mind forms structured 'schemas' from sense experience, though without telling us how. In the late nineteenth and twentieth centuries, the Gestalt psychologists followed this idea in proposing that perception and cognition function by abstracting and reworking the structure in experience. Noam Chomsky incorporated the idea of 'deep structures' into his model of human language. This followed earlier works of linguist Ferdinand de Saussure and of sociologist Claude Lévi–Strauss, who tried to describe deep structures in everyday human activities.

This awareness of deeper structure in experience, and its importance for complex mental functions, thus formed the movement in the twentieth century known as 'structuralism'. Without being sure of its identity, scholars were well aware of something 'there', underlying everyday experience and giving it meaning. The developmental psychologist and epistemologist Jean Piaget spoke of the intuitive 'ideal of intelligibility' that is common to structuralism in a wide variety of academic disciplines, from philosophy to linguistics.[9] The common problem has been that of adequate description (and, therefore, understanding). Usually the awareness of structure has consisted of little more than an intuition of some deeper organisation in some aspect of nature, among its surface elements, without clear or agreed description of what it really is. We see this problem strikingly in structured artefacts: although we all have a compelling notion of structure in music; yet it has defied all attempts at mathematical and scientific explanation.

Describing structure

So how do we describe this complexity, structure or organisation? According to the Oxford English Dictionary, structure consists of 'the arrangement of, and relations between, the parts of something complex...the quality of being well organised'. Likewise, we can find organisation defined as 'the structure or arrangement of related or connected items', or just 'something orderly'. Such definition refers to more than the random composition of a rock, a pile of stones or loose collection of objects. A slightly more analytical account might refer to parts or components that interrelate, are in harmony, are interdependent, make up a whole that is greater than the sum of its parts and so on. But, we might ask, what kind of arrangements, relations and organisation? What are we *really* seeing when we sense structure? Whatever the difficulty of description, we usually feel sure that the structure is 'there' to be comprehended.

In some cases the structure may be quite apparent, just as we are all acutely conscious of patterns in nature, such as frost patterns on window panes.[10] We can also be aware that different aspects, or variables, in nature may change together, or that, change in one may actually cause change in another: we say that they are associated. Such associations are central to structure and can be extend well beyond two aspects or variables. There is an obvious association between daylight and time of day. But increasing daylight is also associated with increasing temperature (and much else). The presence of one entity, such as a specific odour may, to a predator, be associated with nearby prey.

Such statistical association or correlation is the simplest kind of structure. By such structure the entity is organised and perceived as having some degree of complexity. More importantly, this structure offers predictability about the world that can be crucial for the survival of living things: when structure is perceived so is predictability. Increasing daylight predicts warmth and more food a little later (and so time to start producing offspring, or other behaviours). A certain odour can predict the presence of a predator (and time to move elsewhere). These associations may vary in strength, in that values on one variable may be closely, or more loosely, related to those on another. So the predictability may be weak or strong. But, generally, any system evolved within living things that can use such associations to make predictions about the world has also developed a kind of knowledge-abstracting process, and acquired knowledge of the world. That would be an intelligent system.

There is, of course, much more structure than that in the world. There may be multiple associations between variables (as in time of day, light, temperature). And the associations themselves may be conditioned by (or 'depend upon') the values of one or more other variables. For example, the association between daylight and temperature may depend upon season. The association between volume and weight of a liquid will depend upon the liquid (water or mercury, for example). In statistical mathematics this 'deeper' relationship is described as an interaction. Such interactions, too, may vary in strength. Finally, these associations may be 'non-linear': therefore, instead of one-to-one associations, values on one variable may be associated with uneven values in another, as in Figure 2.1. Add in time as a variable and the structure becomes a dynamic one – a merely spatial structure becomes spatio-temporal. This introduces vastly richer structures, just as those in dancing figures and waterfalls are far more interesting than static snapshots of them.

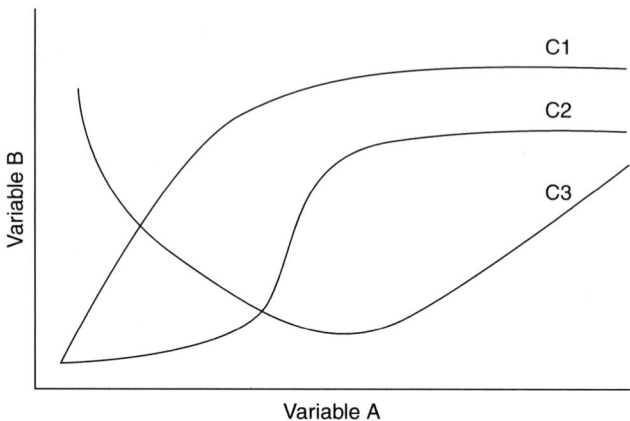

Figure 2.1 Graph of interactive and non-linear relationships. The non-linear associations between variables A and B is conditioned by (varies with) values of variable C (C1-C3)

Of course, the set of non-linear, interactive relationships just described can themselves be conditioned by a fourth variable (a second-order interaction), a fifth variable (third-order interaction) and so on: the structure becomes increasingly complex. We can just about describe interactions at the depth shown in Figure 2.1 verbally, and even picture them visually. But beyond that, at least without considerable practice, the structure tends to be beyond our immediate grasp, especially where it involves non-linear associations. Now throw in a temporal factor, through which any of the above variables, or their relationships, change over time, and we start to get some very complex structures indeed.

A very important point arises from this increasing complexity, though: variables that interact can be highly informative for predictability, in a way that independent variables or cues cannot. Variables such as rainfall and average human body weight in Britain may appear to be 'independent' and do not inform or contain information about each other (indeed, there is no structure there at all). In additive relationships, on the one hand, you 'get what you see'. Interactive relationships, on the other hand, contain deeper information: new or hidden form can be contained, along with more predictability.

Therefore, an organism that has registered the structure in Figure 2.1 can theoretically predict the value of one variable from given values of the other two, with better precision than if they had been independent variables. In statistical modelling the structure is coded as interaction 'parameters': for example, one parameter will indicate the value or strength of the relationship between A and C; another will describe that between A and C as conditioned by C and so on. Later, in Chapter 7, we will see how networks of neural connections in the brain seem to be designed to capture just such relations, and have even been described at the biochemical level in those connections.

An important point is that change over time boosts that information, and associated predictability, through the deeper structure of the change. Compare a flowing river or waterfall with ones frozen, dancing figures with a static snapshot, or a stream of music with a momentary blast of notes. There is structure in the static (constant) version: it's apparent on the surface. But real life is more like the flowing dynamic versions where the structure is more complex and deep. In those cases, a richer structure – and, with it, predictability – becomes manifest in the course of change over time: we can predict a future state from the state now. That is, it is a *dynamic* structure. As researchers have looked more closely at change over time they have found complexities and depths of structure in the world not previously imagined.

Already we begin to see that, in dealing with this structure, living things are dealing with something other than the surface presentation of physical and chemical entities: they are dealing with *information*. What they, and we, see, in structure is information-for-predictability. Unlike total disorder, this structure furnishes possibilities for fruitful response or adaptation. 'Information', however, is another idea that requires some reflection, having been used over the last few decades in more or less vague ways.

Used more technically, the term 'information' is a reflection of the degree of correlational structure in entities, as just described. This is the degree to which values on one variable can predict values on one or more others, and is sometimes called 'mutual information'. Perhaps not surprisingly, such statistical structure has been proposed as a measure of the degree of complexity of a system, and, more widely, as a basic concept for research into complex systems.[11] Informational structure like that becomes meaningful to an organism when the latter has some way of registering it in order to predict some other or future state from it. This may include predicting the consequences of some course of action (e.g., swim to the light for warmth/food).

It is crucial to recognise that living things adapt to these even slightly more complex situations *through* these more abstract informational processes. As we will see, it is such processes that govern the flows of energy in natural phenomena, relations between molecules in metabolic networks, the content of sensory processes in eyes and ears, communication between networks of nerve cells, relations between individuals in societies and so on. The laws that govern them are both derived from, yet *more than*, the basic laws of physics or chemistry. This explains why 'biology cannot be reduced to physics'.[12] Later we will find similar reasons to explain why the laws of cognitive systems cannot be reduced to the laws of biology – and why human sociocognitive functions cannot be reduced to laws of primate (or other species') cognition.

From determinism to dynamics: the new complexity

Great strides have obviously been made in the history of modern science by exploring and describing relationships between variables like these in more or less complex situations. Until recently, however, descriptions were confined, by and large, to linear relationships across a narrow range of values in a narrow range of variables or aspects. The approach goes hand in hand with the mechanistic, deterministic view of the world mentioned earlier – a view in which it is assumed that, given initial conditions, we could always, in principle, predict what the outcome of a change or intervention would be, and we can thereby control the world.

This was the mechanical view of the world that descended on all of science in the eighteenth century. Descartes, Huygens and others tended to assume that all of matter, living and non-living, could be described in terms of simple mechanical systems, such as the swinging pendulum or the coordinated cog wheels of the clock. As Ian Stewart says, it is a point of view in which 'the universe appears to function as a vast system of superimposed cycles'.[13] With such a view a formidable sense of potential power over nature developed. Eventually, in the writings of La Mettrie, Thomas Hobbes and others, the approach became the paradigm for the study of minds and society, too, and, in the mid-nineteenth century, brought biology and psychology firmly into this deterministic world.

The approach tended to look at variables and changes in the world, within confined limits, as just mentioned: that is, linear associations and narrow ranges of values on few variables. These were envisaged to exist, moreover, in an 'equilibrium' state, such that a change in one variable would have responses in others such as to restore that state. What happens beyond these limitations is now turning out to be quite a different story. Advances in physics and other studies in the twentieth century have been discovering that linear, equilibrium relationships do not hold true for the great majority of real world situations. Most involve more numerous variables in non-linear, and deeper, relationships.

It was the French mathematician Henri Poincaré who discovered, early in the twentieth century, that fundamental changes in structure, and advent of more complex structures, can arise even in seemingly simple, non-linear systems such as some fluids and chemical mixtures. Environmental changes affecting variables, perhaps just slightly beyond the range usually experienced, can shift the system beyond an existing region of equilibrium, with surprising outcomes. This suggests a lot more going on than what linear, deterministic models suggest. We now know that natural systems, living or non-living, are very often made up of numerous, perhaps multitudes, of interacting variables.

Accordingly, 'the linear 20th century model' has been subjected to increasing criticism. R. Mateos and his colleagues have summarised some of these as follows[14]:

1. An excessively reduced vision of the dynamics of the real world, and of the origins of structure within it.
2. A simplification of reality by only considering it in a current, local environment, under ideal conditions.
3. Complexity in linear models is envisaged only in terms of increased numbers of variables, and any unpredicted or irregular behaviour is attributed to random statistical noise.

Since the 1950's the availability of large, high speed computers has greatly advanced these discoveries. Huge bodies of data, with many values on many variables, can now be analysed, often with very surprising results. Further investigations have revealed the extent of non-linear interactions in natural systems. Over time, feedback can often arise between the outputs of a system and the operations of the system. In a chemical reaction, the product may activate or inhibit the rate of its own production. Within bodily cells, changes in the three-dimensional shape of enzymes, due to factors such as temperature or salt concentration can transform their roles in metabolic processes from catalytic (promotional) to inhibitory. These all add complexity to a system.

Classical science, however, tended to avoid such complexity: it reduced natural entities to as few variables as possible, sequestered from their natural contexts and pretended those to be linear and time-free. The new approaches examine the behaviour of great numbers of variables simultaneously, in their natural contexts, including non-linear relationships and changes over time. They have shown how multivariate, non-linear systems have properties quite different from simple linear ones. For example, they can respond in unexpected ways to external or internal perturbations, converting a previously homeostatic, deterministic system into one whose behaviour is distinctly different from the original.

The consequences of such effects, 'far from equilibrium', depend on both the nature of the disturbance and the susceptibility of the system. We are all familiar with the sharp, well-defined transformation from one state of order to another, as when ice turns to water. But numerous other kinds of transition can arise in other systems, often reorganising the whole system in previously unpredicted ways to produce new structures, and new complexity, obeying different 'rules'. Moreover, the effects of *that* change, on both the environment and neighbouring systems, can create chains of processes of increasing complexity.[15]

Little wonder Karl Mainzer warns us that 'linear thinking may be dangerous in nonlinear complex reality.'[16] Greg van Orden and colleagues put it more strongly when they say the linear assumption in scientific models, as in popular thinking, 'is too often a specious assumption about complexity in nature, especially the complex behavior of living beings.'[17] In sum, we now begin to see how, within the interactions of variables under changing conditions, complexity may begin to arise naturally, from apparent simplicity, without the intervention of an external designing hand. A static, deterministic view of complexity has given way to a dynamic one.

Attractors and chaos

Of course, in dealing with huge numbers of interactive variables, dynamicists have had to introduce new methods of description, as well as new concepts. The standard way for dynamicists to describe dynamic structures has been to use series of differential equations. An alternative kind of description reduces complex dynamic structure to a more visual description called the 'phase diagram'. Attractors, or the states towards which evolving dynamical systems tend to converge, are said to exist in 'phase space'. Generally, we have to think of a system perambulating among its various possible states – that is, combinations of variable values – until finding the one or few into which it finds equilibrium or is 'settled'. This is the 'basin of attraction'. The phase diagram is the same multi-dimensional space compressed into just two or three variables, duly plotted on paper. This gives us some idea of the dynamic behaviour of the system over time.

In this way, various kinds of attractors have been identified in natural systems. Some dynamic systems tend towards 'steady-state' or 'point' attractors in which nothing much seems to happen, such as a pendulum at rest. A second type of attractor is the 'limit cycle' attractor in which we observe the system reacting to perturbation such as to produce repeating cycles within a limited range of variation. The classic example is a hypothetical pendulum swinging free from other variables such as friction and air resistance: the archetypal equilibrium system (Figure 2.2).

In some dynamic systems the attractor is not represented by either of these and takes on a more unusual, nonregular shape in phase diagrams. They are called chaotic attractors, with interesting properties. To the general public, and in dictionaries, chaos is disorder, turmoil, turbulence or undesired randomness produced by unfortunate disturbance: anything but structure. But, to dynamicists, the term chaos referred to the way that effects of a perturbation can be way out of proportion to the magnitude of the original impulse (sometimes called the 'butterfly effect', from the idea that the flutter of a wing in Brazil might result in a tropical storm in the Gulf of Mexico). This alone suggests more hidden structure than independent variables. One of the better known chaotic attractors – compressed so that it can be plotted in three dimensions – is shown in Figure 2.3.

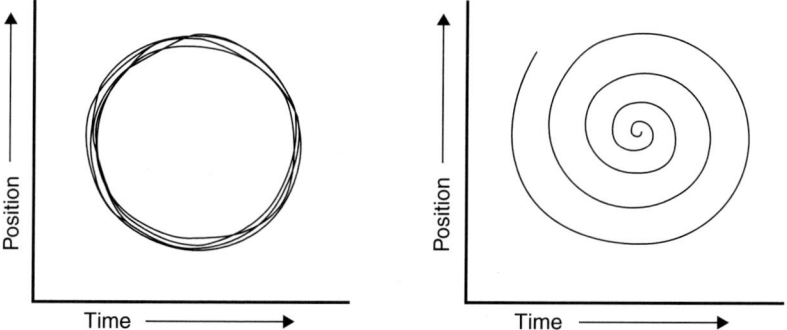

Figure 2.2 Phase diagrams showing movements of a frictionless pendulum in a limit-cycle attractor (left) and a normal pendulum settling into a steady-state (or point) attractor (right)

Figure 2.3 The Lorenz attractor is a graphic description of patterns of behaviour of complex systems like the weather

Source: From *Wikipedia, The Free Encyclopedia*, accessed May 7, 2006. http://en.wikipedia.org/w/index.php?title=Lorenz_attractor&oldid=50807663.

Indeed, the modern idea of chaos largely arose through the discovery of the way that tiny changes in initial conditions can, in some systems (such as the weather), totally alter outcomes some time later. Rapidly changing, or novel, conditions may drive an apparently equilibrium system to a state of instability, sometimes called the 'edge of chaos' or 'criticality'. In this far-from-equilibrium state, the chaotic attractor cycles rapidly through all permutations of variable values, searching, as it were, for optimum, or 'least energy', resolution to the perturbation. New structure and organisation may then appear with a new preferred state (i.e., basin of attraction). This organisation arises spontaneously (is self-organised), purely through the way that energy is dispersed (i.e., as efficiently as possible). All the structures in the world we observe today have their origins in the 'primordial non-equilibrium' of the early universe when changes such as cooling and heat dissipation created non-equilibrium states and self-organized emergence of new structures.[18]

Of course, it is in the nature of complex, interactive systems that the information within them is organised hierarchically – one set of relationships conditioned by another, and so on across numerous levels. This is what is meant by 'deep' structure. Accordingly, another way of describing complex structure has been in terms of correlation across different levels of analysis. In natural structures such as snow flakes or tree branches, smaller structures are 'nested' within bigger structures with some recurring, self-similar (i.e., correlated) features across levels. Measuring the coastline of Great Britain with a long stick will give a shorter distance than the same distance measured with a shorter stick, because the latter includes more detailed features. Yet the two sets of features nest within each other, and will exhibit some correlation of pattern. Similarly, if we were to magnify one of the lines in the Lorenz attractor, above, we would see it was not just one line, but many. Continual magnification yields the same result, hence the shape is said to be self-similar.

This complexity of structure, across numerous levels, is often referred to as 'fractal' structure (from the Greek 'fractus', or broken). Again we have a deeper structure that can be described graphically, or mathematically, in which informational (correlational) relationships define structure and complexity. In natural (living or non-living) structures there can be such correlative associations across many levels, especially when spatiotemporal information is considered. Mathematical techniques and computer modelling reveal fractal structure in a wide variety of both physical and living systems (including the brain), and lead to inferences about hidden structure. Indeed, the deeper harmony and predictability that is found in natural, complex structures, is reflected in the way that people express an aesthetic preference for them over non-fractal structures.[19]

Over the last twenty years, more complex, and formidably sounding, mathematical techniques such as 'Fourier analysis', 'Lyaponov exponents' and 'Recurrence Plots' have been used to describe the deeper structure in complex systems. However, if we bear in mind that the natural world is steeped in self-organised complexity, that can increase with change over time, often with novel 'emergent' forms, we get the essential points.

Note that, even in chaotic systems, the process may still have an underlying predictability that can be potentially learned (e.g., by scientists), or that a living system can utilise for predictability. It is this possibility that gave rise to the hope that the dynamics of complex systems could be, in principle, understood and described by conventional (deterministic) mathematical techniques. For that reason it is sometimes referred to as 'deterministic chaos'. To recent dynamic systems scientists, then, chaos is a theoretically tractable process sensitive to initial conditions, in which an outcome is dependent on structural relations among variables but, in reality, impossible to predict with perfect precision because all initial conditions cannot be known with any precision.

All of this has many implications for understanding the environments to which living things are said to adapt (by more or less complex means), and for understanding the origins of complexity in living things themselves. It portrays a picture of natural structure quite different from simple on–off perturbations: and of complexity as different from mere heterogeneity in these. Most crucially, a dynamic systems approach acknowledges non-linearities and interactions that include time as a variable. The deeper, largely unseen, structure within is the essence of predictability, rather than something of a nuisance. The study of complex systems thus focuses on these dynamic structural relations between variables. It is that kind of deep structure that furnishes properties shared by all complex systems, living and non-living. These include: self-organisation over time and space; emergence of novel properties; adaptability to changing conditions and, in living systems, autopoiesis (or self regenerating structures and functions).[20]

Self-organisation and emergence of complexity

These ideas also help us to understand the *origins* of complexity, including life itself. The mathematician Alan Turing first proposed that patterns of many kinds – animal coat patterns, butterfly wings, ladybird jackets – could form by natural physico-chemical processes, and this idea has now been adequately supported.[21] Ilya Prigogine, the 1977 Nobel Prize winner in chemistry, pioneered studies in relations between systems (such as chemical reactions or living cells) and their environments. He studied systems open to perturbations from surroundings. These he called 'dissipative' systems in that sources of energy enter the system, get used in reactions, and the wastes flow away (dissipate). Prigogine showed in such systems that more complex structures can evolve from simpler ones, and order can emerge out of disorder, by virtue of energy flows finding lines of least resistance. Nearly all systems experience continual energy and material interchange with the environment or with other systems. As a result, change is inevitable, with effects that can evoke new kinds of structure and behaviour from the system, defying assumptions of stable equilibria.[22]

For example, a layer of liquid between two plates will quickly absorb a small heat perturbation and then return to its homogeneous, equilibrium condition. But, if persistently heated from below, it eventually exceeds a critical point at which the liquid forms a new kind of motion. This movement is not random, however, as it can be seen on closer inspection to consist of closely packed and structured convection cells moving in alternate directions (known as Bénard cells, after the person who studied them) (Figure 2.4). The interactions between moving

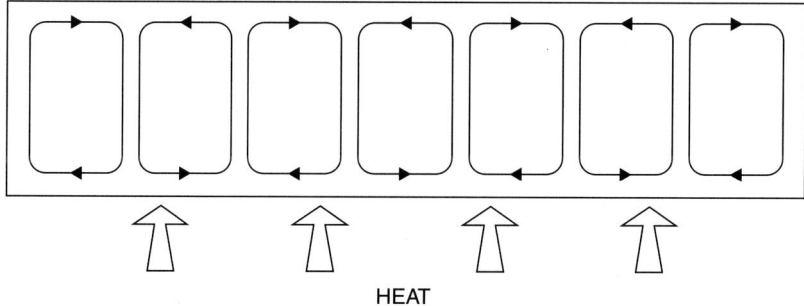

Figure 2.4 Bénard cells forming in a layer of liquid

molecules has lead to a coordination of movement as the most energetically economic form of behaviour, and a more efficient means of dispersing heat. Structure and complexity have emerged without the intervention of an external coordinator.

Over the last two decades there have been demonstrations of hugely complex structures, such as spirals, coils, helical clusters, sometimes arranged as hierarchies, self-organising from chemicals in apparently inert liquids.[23] The physical world is full of such ordered complexity, from the six-sided structure of snow flakes to the crystals of many minerals. More dynamic examples include the turbulence patterns of liquids and gases under stress, tornado vortices and weather changes. The world as experienced mostly consists of dynamic systems. The river we may be gazing at is a flow of water molecules and these molecules, through the structure of fluid forces, self-organise into waves, whirls, eddies and currents.

But these aren't, of course, living systems; nor does this recognition tell us how we get from there to living structures. Nevertheless, such complexity is found in all situations living systems have to adapt to, and, correspondingly, in the systems they adapt *with*. Not surprisingly, then, emergent structure is especially common in living systems where so many interactions are non-linear: for example, the self-organized structures of honeycombs, bird flocks, animal societies, traffic flows and epidemics. In each case a new organising factor now prevails. As Stewart says, 'This is a far cry from Laplace's triumphant determinism.'[24]

As we will see in greater detail, later, such properties furnish living systems with much greater adaptability, or capacity for changing themselves, in changing circumstances. The effects of perturbation are such as to 'stretch' a system to a critical level at which sensitivity and responsiveness to circumstances are maximised. This is 'criticality', a key property of self-organised systems, now known to be common in biological systems.[25] Such dynamics self-organise as systems offering far more 'creativity', or response options, than might have been envisaged through our old reductionist spectacles.

These dynamical ideas are decisively changing views of the world and of the nature of living systems adapted to them. Out of chaos has come the self-organised changes, and emergence of new properties, that have genuinely surprised and delighted scientists in the last decade. But we need to be prepared for it. As James Crutchfield says, in a passage strongly applicable to psychology, '[m]uch of our appreciation of nature depends on whether our minds... are prepared to discern its intricacies. When confronted by a phenomenon for which we are ill-prepared, we often simply fail to see it, though we may be looking directly at it.'[26]

The really significant aspect of these discoveries is that the assumption of unchanging order which animals and minds are durably adapted to is wrong because complex order itself comes from change. This is why I think there have been so many problems in describing a convincing chain of evolution of complex systems. Appreciating natural complexity has radical implications for understanding the true nature of complexities in living things, especially those in intelligent functions. In the next chapter I will describe how it helps us understand even the origins of life. Later, I will offer several illustrations to show that even very basic living systems are intelligent, information- and knowledge-using, systems.

Finally, though, let me attempt to summarise what we should have learnt from this chapter.

1. Complexity is best described in terms of depth of structure, which means numbers of variables involved in a system, and the depths of interaction between them – the deeper the structure the greater the complexity.
2. Natural phenomena are not merely as they appear on the surface, but have deeper structure.
3. That structure, when involving non-linear relationships, and change over time, can self-organise into novel structures with the emergence of novel properties.
4. Differences in depth of structure distinguish the more from the less complex environments.

5. Living things can adapt to complex changing environments if they can somehow assimilate that underlying structure and the predictability it affords.
6. The depths of environmental structure they can assimilate describes the more from the less complex living systems (and what most of the rest of this book is about).
7. The internal, multivariate, complexity of organisms means that their metabolic, physiological and (as we shall see) their cognitive processes obey non-linear dynamics, often resulting in novel solutions to environmental challenges.

3
Intelligent Life

One thing we learn from the previous chapter is that the environment is a lot more complex than we have usually imagined – and certainly more than the standard Darwinian model of adaptation has imagined. But it is often complexity that living things have come to cope with, because it is complexity with structure. Without that there would simply be no life. The structure is particularly rich and deep – in a mathematical sense, but also in an aesthetic sense – as the world changes in space and time. This spatiotemporal structure (changing together over space and time) affords living things a purchase of predictability on the world. If they can somehow learn or abstract that structure, or some of it, then they can anticipate the future from the present, respond in fruitful ways, and eke out a living.

Because of that possibility, the existence of such structure explains a richer evolutionary story then we have usually been told. The 'simple ways of life' will tend to get used up (as Darwin said). They might become overcrowded, and force overspill, or just become more complex through dynamic interactions within. Either way, those varieties with any propensity for more complex traits (perhaps ones redundant or precocious in the past) now find themselves at an advantage. They will tend to survive and leave more offspring (the complexity is naturally selected). So complexity evolves.

But having to cope with changing, structured environments changes the game fundamentally. No more the fixed, recurring environment to which bodily structures can adapt by genetic selection over many generations, culminating in a nice lock-and-key adaptation. We now have dynamic environments, changing over seconds, hours, days, weeks or years, within the time spans of single generations. This means evolving different *kinds* of characters or traits. Instead of fixed features, organisms need functions that can abstract predictability from environmental structure after birth and throughout life. Changing environments that have deeper structure, demand systems that can detect and engage with that structure – systems that are rather more complex than simple cue-response adaptations. This is what we mean by *intelligent* systems.

Dynamic environmental structure is what favoured the evolution of more and more complex systems, and of increasingly integrated organisms, and minds. This chapter gives some examples of the structure-abstracting devices that pre-evolved cognition, before we move on, in later chapters, to more complex cognitive systems themselves. The first link – and, perhaps, the biggest obstacle to understanding – is that first amazing leap in complexity, the origin of life itself. Basically, I will try to show that life is part of the world pinched off, and constantly striving to maintain its integrity in relation to its parent – but one in which the parent is constantly 'upping' conditions of its offspring's survival. It is that relationship that has created the sequence of bridges to complexity in living things.

Origins of life

We all recognise that there's something more complex and different about living compared with non-living things. School and college texts still define 'life' in terms of criteria such as respiration, ingestion, excretion and reproduction. But these are just lists of properties that don't seem to quite reveal the essence of life. Nor do they convey a sense of life's complexity. Even the simplest of living organisms consist of thousands of different specialised molecules, interacting within elaborate webs of chemical processes, sensing and signaling about the surrounding world and responding in ways such as to maintain their internal cohesion. Even today, scientists, as well as ordinary people, wonder how undirected, unmanaged, physical forces could have produced such immensely complex states from a few simple ingredients. So, even quite recently, Carol Cleland and Christopher Chyba could argue that biologists still have no theoretical explanation for the origins of life, and that, '[i]n the absence of an analogous theory of the nature of living systems, interminable controversy over the definition of life is inescapable.'[1]

Astrophysicists are now fairly sure that the earth originated between 12 and 15 billion years ago (though how is still open to debate). The appearance of life on earth some 4 billion or so years ago probably depended on the earth's crust having cooled sufficiently for liquid water to exist. But how did this momentous appearance occur? In their book, *The Origins of Life*, John Maynard Smith and Eors Szathmary describe life as following a number of transitions, starting with molecular systems that can replicate themselves, and ending with human societies, including language and mind.[2] This is a fairly conventional picture. But what has driven this process? How did we get increasingly complex entities with such original properties from structureless matter?

Genesis in genes?

A widely held argument is that genes are the codes for living structures, so something like genes must have been there at the start to design life systems. This follows from the enormous – indeed, quasi-cognitive – powers that have come to be imputed to genes as 'designers' or 'controllers' of the development of organisms from birth. Various suggestions have been made about where *genes* came from, including Francis Crick's idea that they came from outer space (which, of course, only displaces the problem).[3] However, thanks to Crick and Watson we know that the chemicals of the genes, DNA and/or RNA (see more on these below), consist of precise, but variable, sequences of subunits called nucleotides, held together on a sugar-phosphate spine. These sequences, indeed, act as molecular 'templates' for the construction (via intermediaries) of the complementary sequences of amino acids in proteins, the building blocks of living things. These templates can also be copied directly into identical strings of nucleotides to form copies of themselves to be passed on to offspring cells. They thus seem prime candidates as the basic 'molecules of life' and for explaining where life came from.

The problem is that the 'genes first' theory presents a classic chicken and egg conundrum. We now know that reading of instructions from genetic templates for the production of the thousands of ingredients that would then 'spring' into life is far too much to expect. It would require, *in advance*, hugely complex supply chains of ingredients, as well as the coordination of a host of chemical reactions, to bring about faithful transcription and the step-by-step sequencing of proteins, or even self-copies. How could the DNA have gathered about itself all the prerequisites for these processes in advance? Not surprisingly, years of careful effort in laboratories to find processes through which genes can produce chemical components of living systems, and string them together in just the right order, at just the right time, without those prerequisites, have failed.[4]

Life before genes

The logical conclusion is that the distinctive qualities of life were not originally based on genes, but rather molecular ensembles.[5] Alternative origins-of-life scenarios – what we might call 'genes last' views – have thus become popular. Rather than life having its origins in a pre-existing genetic blueprint, other arrangements have been envisaged. At the time of writing his *Origin*, Darwin knew nothing about genes, but how that first living complexity came about greatly troubled him. He suggested, in the final section of his great work, that 'the Creator' must have originally breathed life 'into a few forms or into one'. Evolution by natural selection – accumulation of small advantages – then took over. 'From so simple a beginning', he went on, 'endless forms most beautiful and most wonderful have been, and are being evolved.' Later, in private correspondence, he expressed doubts about supernatural intervention, and suggested life could have arisen from chemical reactions, 'in some warm little pond, with all sorts of ammonia and phosphoric salts, light, heat, electricity, etc. present.'[6]

Darwin's conjecture, ironically, may not have been far from the truth. It is consistent with the many demonstrations of self-organisation, mentioned in Chapter 2, and became, through theorists such as Haldane and Oparin in the 1920s, the 'primordial soup' theory.[7] Then, in a historic set of experiments in the 1950s, Stanley Miller at the University of Chicago passed electrical discharges through mixtures of hydrogen, methane, ammonia and water vapour in the laboratory, as if to simulate lightning passing through a primitive earth atmosphere. In just a few days, more than 15% of the mixture had been converted to a variety of amino acids, the building blocks of proteins and other 'organic' molecules, as potential biological constituents.[8] These results have recently been reanalysed to suggest an even wider range of products than originally reported.[9]

There have been doubts about Miller's conception of early earth's atmosphere. Besides, it now seems more likely that the early organic ingredients were formed from the catalytic properties of the surface of metals, such as iron and nickel, on the floors of hot, acidic oceans using the abundance of hydrogen and carbon gases.[10] Either way, not only can amino acids so form, but they can also readily combine into strings (polymers), the basic structures of peptides and proteins.[11] Many of these strings are now known to have at least weak catalytic properties, assisting the formation or breakdown of other molecules to form a more complex system of components. These could interact with each other and with others from nearby to form so-called autocatalytic sets. Within these sets, the product of a chemical reaction catalyses one of the reagents, accelerating the formation of product.

Some investigators have suggested that the conditions of the early environment on earth were such as to make origins like this statistically inevitable. Indeed, autocatalytic reactions among likely components were discovered in 1996.[12] We now know of a variety of polymers and small molecules that can catalyse such reactions.[13] Among these are nucleic acids – the chief constituents of DNA and RNA – that can spontaneously polymerise into strings and also catalyse the formation of amino acid strings (the building bricks of proteins). Doron Lancet's group envisage a process in which organic molecules 'accrete together spontaneously, even from dilute "soup" solutions, to form assemblies, whose dynamic behavior manifests life-like attributes', including reproduction of constituents.[14] They showed in the laboratory how such ensembles are capable of propagating their constituents and of self-reproducing without RNA or DNA 'codes' as such. We see this in many common metabolic cycles in which copies of molecular constituents are made. Only later did the genetic polymers (RNA then DNA), what we now call genes, arrive on the scene.

Such autocatalytic functions (some weak) are known to reside in a wide range of readily available molecules, including certain amino acids, peptides, lipids, carbohydrates and metals. They can create wide reactive networks and may have constituted primordial 'metabolisms': self-sustaining, self-organising forms that take in energy from outside the system to maintain their structures, dissipating wastes back into the environment.[15] Indeed, Stuart Kauffman argues that any large variety of reacting molecules put together is likely to produce at least one autocatalytic cycle.[16] Mutually catalytic networks have what Barak Shenhav and colleagues call 'compositional information'. In these, the *interactional* structure among components determines the product rather than the sequential information we are more used to in genes. Using computer models, they argue that such collections arrived very early in the evolution of life. Accordingly, 'molecular ensembles with high complexity may have arisen, which are best described and analyzed by the tools of Systems Biology. We show that modeled prebiotic, mutually catalytic pathways have network attributes similar to those of present-day living cells'.[17]

In this kind of scenario, then, it seems reasonable to propose the original accretion of organic molecules engaging in such autocatalytic cycles and slowly accommodating additional molecular partners.

The networks emerging would persist by breaking down energy rich molecules ('nutrients') from the surrounding medium to replace those worn or denatured. The by-products could then be dissipated into the medium. Various environments have been envisaged as the cradle of such activities, including ocean floors, deep-sea volcanic vents and so on. But what distinguishes them from non-living collections of interacting molecules is their ability to maintain their integrity across wide ranges of environmental change, at least for some period of time. This is, at least, some sort of life.

Of course, any metabolic network or system will eventually disintegrate, or 'die', due to the cumulative effects of random degradation of molecules. Replacement by wholesale reproduction is then necessary if the system is to continue as a coherent identity. This is why the idea of reproduction is such a crucial criterion of 'being alive'. Self-replication of peptide, protein, lipid and a variety of mineral-based molecules *has* now been demonstrated. In addition, replication of components is known to arise from a number of metabolic cycles, as just mentioned. Self-assembly of complex bio-components such as ribosomes and viruses is also well known.[18] But wholesale copying of all cell components, in a more or less concerted manner, from ingredients happily present in the medium, would seem very improbable.

One idea is that these autocatalytic sets of components may have become split by various possible environmental perturbations (turbulence, temperature or whatever) into two or more similar mixtures. All in all, as Shenhav and colleagues put it, 'the conceptual outcome is highly important.... [M]olecular assemblies which undergo dynamic exchange of matter with the environment, are energy-dependent, grow homeostatically, and have a potential to generate progeny'. Moreover, the process may have been accelerated by the non-linear dynamical quality of the internal interactions among numerous components. 'They may assume stationary states far from equilibrium and display properties, similar, in a rudimentary way, to those of present-day living cells.'[19]

Whatever the details, the 'metabolism-first' approach assumes that self-maintaining, autocatalytic, molecular systems, rather than genes as such, were the seeds of life. The production of polymers that could act as templates for production of components 'in-house', as it were, came later. We know that the formation of RNA strings from component nucleotides is a relatively straightforward event, and that RNA can combine the necessary properties of sequence templating (encoding) and of catalysing chemical reactions. So RNA was probably the first candidate for this role. But the job is now mainly done by DNA, perhaps because of its greater robustness in hostile environments.[20] RNA continues to serve a variety of other functions, though, as we will see later (see Chapter 4).

Life evolves

However plausible this kind of scenario for life's origins may be, without evolution, that other crucial criterion of life, such molecular gatherings would remain as isolated curios among a myriad other interesting physico-chemical systems. All would be involved in self-replicating into mountains of the same relatively boring stuff, and unable to cope with change. How, then, did they evolve into something approximating an identifiable unit, or cell, full of 'life', of increasing diversity and complexity, as we currently recognise it?

In fact, variation can arise in many ways. As just noted, the accommodation of new partners in the metabolic network will automatically create variation: different webs with different reaction properties to varying conditions. Environmental change itself will create variation in reaction speeds among molecular components, and so on. Similarly, environmental perturbations may produce errors ('mutations') in the self-reproduction of molecular species, producing further diversity: for example, the molecules constantly jiggle because of Brownian motion. Above all, the very activity of autocatalytic sets produces waste chemicals that diffuse into, and change, the local environment, requiring continual change within the sets to survive. Such interactions – an early form of co-evolution – may also have been the earliest examples of 'chasing complexity' just to remain 'alive'. This is why continuity of (changing) identity under environmental change is a crucial criterion of life.

In the midst of such diversity, of course, in which a myriad molecular 'experiments' are being tested in changing environments, the first natural selection must have occurred, simply as a logical truism. It is easy to imagine a kind of Darwinian process operating on this variation to favour ensembles with different degrees of robustness in different environmental conditions, and so tracking environmental change. A system that can adjust to changes in even single variables such as ambient temperature or pH has already been 'selected' ahead of a system that fails to adjust.

There is nothing special about the process: it has been demonstrated that Darwinian selection can work in a simple, seemingly lifeless, system of chemical reactants.[21] Many aspects of a molecular ensemble may be subject to a kind of natural selection in this Darwinian sense. Speed of utilisation of energy sources may be one. Efficiency of incorporation of components may be another. Buffering within the system against environmental stress or fluctuation will have been important. All of these are important in cells in today's organisms and will always give rise to diversity in different conditions with or without additional complexity.

However, as we saw in Chapter 2, there is more than that to environments and, therefore, more to what might be selected. Environmental change *can* involve singular variables such as temperature, salinity, pH, turbulence or concentrations of nutrients. And these can be cues for the selection of some response tendencies (such as resistance to temperature changes). In most cases, however, the environment will be more complex in a sense other than just greater numbers of independent components. Rather than independent perturbations, complex associations may form between variables: turbulence may increase with temperature, but in a way conditioned by turbidity (salinity), for example, while levels of dissolved gases will decrease. Changing environments tend to create patterns from underlying structure, as we have seen. This kind of interactive complexity will become increasingly significant as different molecular ensembles themselves come to form parts of the environment, impinging on, and interacting with, *one another*.[22]

As shown in very simple form in the previous chapter (Figure 2.1), there is additional information in such interactive sets. The levels of some environmental variables can provide predictability (information) about the levels of others. When the variables are related to temporal change, the system becomes a more dynamic one. In other words, there is great advantage to be drawn from any ability a molecular ensemble has to attune to that deeper informational structure. It seems plausible that such 'knowledge', or intelligence, about the environment could have been a defining feature of living systems from early evolution. This is one reason why recent studies on single cells adapting to changing environments have led investigators to ask 'can cells think?'[23] Certainly, in the more evolved systems, even of today's single-celled animals, many mechanisms are known through which molecular ensembles can become attuned to environmental structure, and tailor responses accordingly. Some of these are described below and many others are discussed in Chapters 4 and 5.

In Darwinian terms, networks able to adjust to environmental structure, not just singular cues, must have been important targets of selection almost from the start of life. Interestingly, simulation studies have shown how complex living systems, that are more than bundles of simple cue-response reactions, only evolve in 'complex' environments, consisting of interactive, non-linear relations among numerous variables. This, too, suggests that it is *structured* change, rather than the mere presence of certain elements (ingredients) that inspired the first leap to complexity. As Chris Langton put it, 'Rigid, pre-specified, "unnatural" environments foster rigid, predictable, "unlifelike" evolutionary progression.'[24]

However, it would be quite wrong to conceive of such a Darwinian process as being, on the part of the ensemble, a merely passive one. By virtue of their non-linear, interactive variables, some primitive systems could have had the capacity to respond to novel environmental change more creatively, moving the system to new equilibria and more complex states of organisation, or even generating new versions of components. The varieties thus made available for natural selection are, in part, active contributors to the evolutionary process. In that way, natural selection on the 'outside', and self-organised changes on the 'inside', work in tandem: 'self-organisation proposes what natural selection disposes', as David Batten and colleagues put it.[25] Again, though, the process is a logical product of change in the system, interfacing with change in the environment.

From molecules to cells

Much of this evolution of quasi-living systems assumes the next stage in the origins of living systems. This is the move from naked molecular ensembles to encapsulated cells. This process, too, has been demystified in recent years. It has been shown in laboratories how certain proteins can form membranes that act in many ways like those in living cells, permitting the passage of some chemicals but not others. Sidney Fox suggests that these membranes may eventually have assembled into spheres, or protocells, accidentally enveloping self-organising, self-reproducing molecular systems. The general idea has been supported in more recent research.[26] Bounding the system in a membrane permitted better control over its metabolism, including what goes in and out.

Such cells are also prone to 'budding', where part of the membrane pinches off to form a second cell or vesicle. As long as there are sufficient copies of each molecular component in the offspring vesicles, self-replication and reproduction can continue indefinitely, or until the self-replicating structure breaks down. These may have been the first steps in the evolution of the forebears of primitive, single-cell organisms living in a wide range of more or less changeable conditions. In fact a huge variety of primitive organisms, now classified as the Kingdom *Archaea*, have been discovered over the last few decades surviving in extremes of salinity, temperature and acidity (pH). These were most likely the forebears of other Kingdoms such as the bacteria.[27]

So we have the origins of living systems in natural forces. Eventually, other basic features evolved. DNA became firmly installed as the templates for both regulatory and structural proteins – and thus as *resources* for metabolism and development, not their controllers. It became organised into long strands on chromosomes. The chromosomes then became organised into nuclei, itself surrounded by a membrane to afford greater control over access to, and exit from, the nucleus. The basic qualities of life remain the same, however: living systems are 'open' systems, subject to changes in the outside world, and responding creatively in 'far-from-equilibrium' thermodynamics.

In a constantly changing world, physical equilibrium systems, consisting of pre-designed structures, spell stagnation and ultimate extinction. The true 'life force', on the other hand, is the ability in some of those early self-organising systems to survive the challenge of increasingly complex environmental change.

It is worth stressing, of course, that much of this picture is still under intense investigation, with computer simulations augmenting the limited capacities of the test tube.[28] Nevertheless, a closer look at some more evolved systems supports the plausibility of the general principles.

Intelligence in cells

'Learning and memory – abilities associated with a brain or, at the very least, neuronal activity – have been observed in protoplasmic slime, a unicellular organism with multiple nuclei.' So says Philip Ball, with a hint of incredulity.[29] As mentioned in Chapter 2, and above, environments may occasionally be experienced as repetitive cues that can be dealt with by built-in cue-response associations. More often they appear as stimuli that are superficially novel, but parts of a deeper pattern. Ball is referring to experiments by Toshiyuki Nakagaki and colleagues on a slime mould *Physarum polycephalum* (amoeba-like single cells, common in soil). Shocks administered to the amoeba in the laboratory will slow down its normal motion. If the shocks are delivered at regular intervals, however, the organism will 'learn' the underlying pattern. Impending shocks come to be anticipated at the appropriate intervals and the cell slows down in advance. This 'cellular memory', as the researchers call it, persists for some time after the shocks have ceased.[30]

This experiment demonstrates the ability of even simple organisms to deal with changing environments by abstracting the underlying pattern, though the mechanisms in this case are not well understood. There are, however, other cases in which they are. The point made in Chapter 2 was that living systems survive in changing environments, in which they experience 'constant novelty'. They need to be able to register any underlying pattern generating that novelty, if they are to predict outcomes and respond intelligently to it.

Certainly, nearly all cells alive today, whether as single-celled animals, or parts of multicellular species, are exquisitely sensitive to tiny changes in their environments. These include concentrations of chemical attractant and repellent substances around them, or physical stimuli such as light or heat. Changes are detected, moreover, against a wide range of background concentrations or intensities. For example, the common gut bacterium *Eschericia coli* can detect less than 10 molecules of an amino acid in a volume of fluid about equal to that of its own size – or a few drops in a bathtub of water.

Intelligent cells in motion

One of the simplest forms of cell reaction is to physically move to or away from the source of change, depending on whether it is likely to promote survival or threaten it. Most simple species live in water or fluid layers, and those without motile power of their own are at the mercy of currents, in constant, swirling motion and agitation. They can only adjust to change through adjustment within the cell. In extreme conditions many will, in fact, form hard-walled shells and recede into a sessile state – literally shrivelling up to wait for better times.

On the other hand, many single-cell species have evolved systems of motion for getting to or away from significant sources of stimuli. The most studied of these, especially in bacteria and amoeba, is that of motion called chemotaxis. Bacteria, in fact, can sense a vast range of environmental changes, from the concentrations of nutrients and toxins to oxygen levels, pH, osmotic pressure and the intensity and wavelength of light. The actual mechanical systems of motion vary greatly. *E. coli* have propulsive flagella (tiny hair-like extrusions on the surface) that rotate as spirals, a bit like a ship's propellor. Other bacteria achieve a kind of gliding motion on hair-like 'pilli' over secreted slime. Amoeba move through pushed out extrusions of the cell (pseudopods or 'false feet') forging the direction in which the rest of the cell then flows. But the general reaction cycles have turned out to be very similar throughout the living world.

Studies on these signal-reaction systems have also demonstrated that, to any individual cell, the external environment is not a random array of physical and chemical entities. Rather it forms intensity-, or concentration-, gradients constantly changing in space and time. The critical factor is that organisms sense, and move in response to, the *gradient,* not to a predetermined level of sensation acting as a kind of threshold 'trigger'. This is the case whether the stimulant be concentration gradients of chemical substances (nutrients, oxygen and so on) or intensity gradients of physical variables (heat, light and so on). Even the simplest cells, that is, cannot be described as If-Then, cue-response, machines.

The reason for this is that a world experienced in that simple, direct way is of little use to a single cell. For example, a single nutrient molecule colliding with a bacterium cell wall/membrane gives little indication of where it came from, especially as there will be a storm of other molecules colliding with it at the same time. This is why there are no direct communications between sensors and motor system, even in a singe cell. The direction and distance of an attractant or repellent substance can only be distinguished from background noise in one way: through correlations among large numbers of molecular collisions in relation to the layout and curvature of the cell surface. Making sense of that requires an intelligent system, rather than a simple cue-response mechanism.

This logical dependence of behaviour on signal *structure* rather than on simple cues or triggers is reflected, in *E. coli,* in its complex surface monitoring system. Like all such cells, it is contained in a membrane within a protective cell wall. The main lipid basis of the membrane is studded with special receptor and transport proteins evolved to recognise different substances, usually by chemically combining with them (so those substances are called ligands). These membrane components permit ingestion of nutrients, egestion of wastes and so on, while mediating signals about the outside world. So important is this signal-detection apparatus that genome sequencing projects have revealed that membrane proteins represent about a third of the gene products, not just in bacteria, *but also in most organisms.*

Reconstruction of the spatiotemporal 'outside' is achieved, in other words, by thousands of chemical receptors embedded in the cell membrane. It starts when the signal-triggering ligands become attached to their respective receptors. This creates a structural change that is transmitted to other proteins underlying the membrane. These communicate with each other about what has been received and help differentiate and amplify the signal before passing it on to internal metabolic processes through a chain of intermediaries.[31] The spatiotemporal structure of the outside is facilitated in *E. coli* because clusters of receptors are concentrated at one end of the rod-shaped cell (Figure 3.1). Diffusing substances on the outside will contact the curved surface in an order correlated with the direction of their source.

However, this gradient structure sensitivity is probably most facilitated by a process called 'serial ligation', in which one molecule can attach/detach to and from a series of receptors in turn. This, too, produces a sequence of signals, spatially and temporally correlated with the direction of diffusion from the source.[32] The cooperative interaction among receptors themselves, effectively communicating with each other, may also facilitate this correlation monitoring.

But there is still another, temporal, register of the amount of ligand bombarding the surface. This arises because, after a receptor has bound a ligand, and signaled the event, it needs to be 'reset' by another chemical reaction called 'methylation'. The amount of methylation required provides, in effect, a rudimentary form of memory (of how many receptors have been bound). Internally, that allows bacteria to compare their current and immediate past environments, distinguishing, for example,

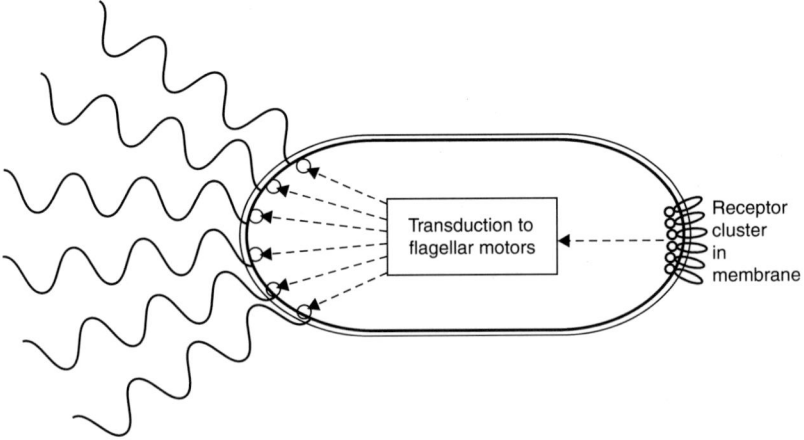

Figure 3.1   Schematic diagram of *E. coli* to show receptors and flagella

between a short pulse and sustained environmental changes, as in a chemical gradient.

Now let us see what the humble bacterium, having internalised this spatiotemporally structured outside, does with it. Motion in *E. coli* is achieved by a rotation of five to eight whip-like extrusions from the cell, called flagella (Figure 3.1). In the absence of an attractant gradient the flagella spiral in a clockwise direction, driving the cell forward in a fixed direction. At random intervals, though, the rotations halt momentarily, causing the cell to undergo a kind of 'tumble'. After a brief rest the flagella motion continues, sending the cell off in a new direction.

This seemingly haphazard pattern of motion continues until an attractant gradient is detected. The structured signal freezes the randomness of tumbling, in effect impelling continued motion in the same direction. A repellent gradient has the opposite effect, increasing the probability of tumbling. The overall effect is for the motion to be preserved in a desirable direction, or for more tumbles and redirections away from an undesirable direction. As a process of way-finding it may sound a little like an intoxicated person finding her or his way home from the bar, in the dark. But it does work. The whole sequence involves numerous proteins in spatiotemporal interactions, and the recruitment of dozens of genes through precisely regulated transcription and further signaling pathways.[33]

In general, then, what may superficially seem to be a system of reflex actions triggered by singular environmental cues is actually an intelligent response to external structure. The sensitivity to structure renders the changing environment, and the consequences of responses to it, more predictable. Similar processes of complex signal transduction and mediation are found in other motor systems, like that in amoeba that 'creep' over a matrix surface, rather than propel themselves with flagella. In fact, the idea that it is spatiotemporally structured changes, rather than absolute values of variables, that matter in dealing with the world, will be a recurring one across all the systems we will be looking at later.

Note, in passing, that these systems of cell motility are of intense biological interest, generally, because they have been incorporated, sometimes with further elaboration, throughout the animal kingdom. In multicellular organisms, many different processes depend on cell locomotion, including migrations during early development, lifelong remodeling of nerve connections, movements of immune cells towards pathogens, fibroblast migration during wound healing and so on.

A bacterium learns

Single-cell organisms can capture temporal correlations in other ways. As Ilias Tagkopoulos and his colleagues explain, 'temporally structured correlations can exist on multiple time scales, reflecting the highly structured (non-random) habitats of free-living organisms. Temporal delays are a typical feature of these correlations. For example, an increase in temperature may herald an impending decrease in O_2 levels some 20 minutes later. An organism that is capable of learning (internalising) these correlations can then exploit them in order to anticipate vital changes in the environment – for example, preparing for resource fluctuations or mounting protective responses to extreme perturbations.'[34] This can happen in single cells, they say, because their biochemical networks can create internal models of the complex environment.

To demonstrate this, they turned to *E. coli* and its responses to environmental change. In normal environments, there is an inverse relationship between temperature and oxygen levels, as just mentioned. Sometimes *E. coli* can be faced with rapid changes in both. When an *E. coli* gets into your mouth (a not infrequent event) it suffers an immediate increase in temperature. Then you swallow, and it arrives in your gut, where there is not a lot of oxygen. Survival in that environment requires a quick switch of energy-releasing respiratory metabolism from aerobic to anaerobic – all of which involves the recruitment of genetic transcriptional and metabolic pathways.

What Tagkopoulos and colleagues showed (by cultivating bacteria exposed to that sequence of conditions) is that the bacterium can use the initial temperature increase as a signal for the impending drop in oxygen levels, so saving a lot of time in making that switch. The speed of transfer to anaerobic metabolism, even when oxygen levels were still high (a superficially maladaptive response) was particularly striking. It suggests that the organism has somehow assimilated the external correlational structure of those events, and is now predicting one from the other.

It might be expected that such adaptability to environmental structure would be somehow due to a genetic program having assimilated such contingencies. The group showed, however, that the apparently 'hard-wired' response can be reprogrammed by cultivating bacteria under the *opposite* regime: starting at low temperature (but with low oxygen) followed by upshift in oxygen. The cells eventually learned the new environmental structure and suitably switched metabolism in advance.

What the authors suggest is a bacterium capable of 'dynamical modeling' of the external environmental structure through the organisation of internal metabolic and catalytic networks. This allows intelligent anticipation of the future and preparation of suitable responses. The demonstration draws attention to the importance of structure – the real source of environmental complexity – over independent elements. As they say, 'inferences regarding the functional utility of biological networks, including notions of modularity and optimality, may be incomplete, or even inaccurate, without considering habitat structure'.[35]

Rhythmic complexity

One source of environmental change that all organisms, including bacteria, have to deal with is that dictated by the rotation of the earth on its axis. The daily rhythm imposes variations on a host of factors: temperature, light, humidity, food availability and many other factors. Depending on locality, the variations can be extreme. Continuous metabolic and behavioural activity, ignoring this variation, would be fruitless for long periods of time, as well as energetically wasteful. Accordingly, attunements to this rhythm have evolved in nearly all species, and circadian (about a day) rhythms of activity are found in nearly all species from bacteria to mammals.

Superficially, it might seem that a simple cue-response function might do the trick: light on → action; light off → inaction. That solution would not be ideal however, because the activity needs to be anticipatory, not merely reactive or coincidental. Even plants need to get their photosynthetic machinery together, with their leaves already lifted, just *prior* to the onset of dawn. In animals, a whole spectrum of functions needs to be 'warmed up' in advance, from sensitisation of chemoreceptors and digestive systems to tensioning of limb muscles. Moreover, the adjustment cannot be the permanent 'gearing' to a consistent pattern that a built-in reflex function would provide. This is because the pattern itself changes with another movement of the earth, the seasonal tilting on its polar axis. So we get different seasons with different lengths of day, also depending on distance from the equator. The environmental context, in other words, is a dynamically structured one. The pattern of light intensity and, with that, a range of other variables (food supply, predator frequency, temperature and so on) are conditioned by the season of the year. As in the other examples described above, such structure permits a predictability that can be assimilated in an intricate set of internal responses and *their* interactions.

Much of the mechanism of assimilation of this structure has been investigated in fruitflies in which lifecycles are closely geared to the circadian cycle. This is because new adults need to emerge from their pupal cases in the morning, when it is relatively cool and moist, to avoid desiccation, and they fly, mate and feed in the day, resting at night. Even pupae kept for a spell in total darkness will emerge at the true dawn so long as they were exposed to a 12 hr light–dark cycle before that. The fly's behavioural pattern indicates anticipation of that in the environment by entrainment of some internal 'pacemaker'. What follows is a mere glimpse of an altogether more intricate process with many more components, and is offered only for general impression of its complexity (so don't worry if you don't grasp it all at once).[36]

The first clues to mechanism arose from experiments in which genes were experimentally mutated (for example, by bombarding with ultraviolet light). It was observed that mutations of a specific gene, which came to be called the *period* or *per* gene, disrupted the behavioural rhythm. Molecular studies showed that the protein product (PER) is transcribed from *per* in a circadian cycle, with the length of the cycle (i.e., more or less than 24 hours) that depends on PER concentration in brain tissue. Higher doses of PER decreased period length; lower doses increased it. However, the *per* gene was found to work in tandem with another gene, timeless (*tim*) and its product (TIM). But that isn't all. Transcription of *per* and *tim* into their protein products is also promoted by the protein products (dCLK and CYC) of two other genes.

Critically, both PER and TIM accumulate outside the nucleus in the cytoplasm of cells during the day, reaching peak levels in early evening. Being in the cytoplasm makes PER susceptible to degradation by the product (DBT) of another gene, *doubletime,* which combines with it – except that the accumulating TIM also combines with PER to prevent that degradation (see Figure 3.2). In the evening, the DBT/PER/TIM complex moves into the nucleus, under the influence of another gene (*shaggy*) and another step begins.

Once in the nucleus, TIM is slowly released from the complex and represses the function of dCLK and CYC. This negates the latters' promotion of the transcription of *per* and *tim*, so diminishing production of PER and TIM. In addition, thanks to this release of TIM from the complex, 'exposed' PER can now be degraded by DBT, which happens over a period of 8–10 hours. The cycle starts up again because (a) TIM gets used up over the course of the night, and (b) PER, which represses its own production from *per*, degrades so that its concentration diminishes. In these ways transcription of *per* and *tim* is again

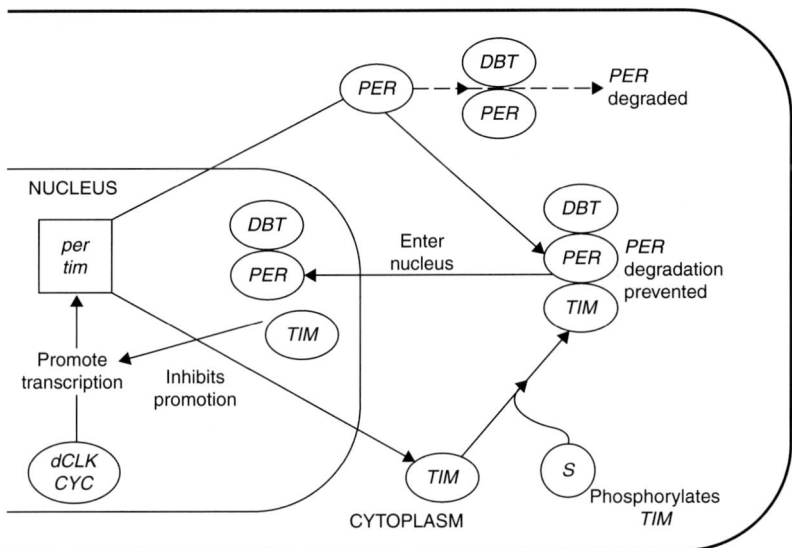

Figure 3.2 Simplified diagram of regulations in circadian rhythm of the fruitfly

activated. With the inhibition lifted the cycle starts again, just in time for dawn!

This already describes a process of adaptation to a structured change, involving interactions among many components. By assimilation of the external structure into the web of regulations just outlined the organism can anticpate the changes and be ready for them. As complicated as this process is, however, it does not explain how the 'clock' is set to anticipate the dawn and dusk as they change with the seasons. Nesting of the cycle in this additional structure requires regulation at another level, involving a light-sensitive protein called CRY. CRY interacts with TIM and degrades it in response to a light pulse (remember that TIM inhibits the transcription of *per* and *tim*). This light-dependent degradation thus alters the phasing of the whole cycle because the decline of TIM advances the cycle described above (the inhibition is relieved). At dusk, TIM protein levels rise, which in turn delays the molecular cycle (it inhibits the promotion of *per* and *tim* transcription). In this way, animals can attune to this deeper environmental structure, and adjust their daily rhythms to seasonal change.

In essence, this system of structural assimilation is similar across 600 million years of evolution from flies to humans, albeit with some modifications. In humans it is even more complicated, other 'clock' proteins having been discovered, and involving feedback from ancilliary pacemakers in most peripheral tissues.[37] Also, other structure in the environment may come into play. Although the light period (photoperiod) is the dominant factor, it is not an independent variable in experience, being integrated with a host of other variables such as temperature, rainfall, food availability, changes of vegetation and activity of other animals. Studies have shown that the structure of interaction among such variables can override the effects of light alone. For example, it was shown in one experiment in Madeira that canaries, normally breeding in spring, and switching off reproductive behaviour in autumn, could start breeding again in December if green plants are placed in the environment.[38] Circadian rhythms in mice, which have been suppressed through exposure to constant light, can be reset by exposure to constant darkness. But the process seems to be regulated by change in behavioural activity around them, rather than by just light or dark alone.[39]

The system is another illustration of how interactions among numerous components can behave with dynamic precision, with widespread consequences. Recent analyses of the joint expression of thousands of genes have revealed hundreds of transcripts that fluctuate with the circadian clock, through neurohormonal signaling systems (including the well-known melatonin). Some of these transcripts have been identified as key regulators/coordinators of sensory, digestive, endocrine and other metabolic and motor activity, and are implicated in obesity, sensitisation to cocaine, cancer susceptibility and responses to chemotherapeutic agents.[40] As Franck Delauney and Vincent Laudet say, 'The important message from these observations is that almost every biological process in a cell or an organism seems to be affected at some level by the circadian clock' and 'every gene in the genome could be under circadian regulation in one organ/cell type or another'.[41]

So here we have an intelligent biological system affecting virtually the whole biosystem from bacteria to humans. But what is its origin? It involves genes. And it involves the environment, obviously. But not by any of these independently. It is regulated by a dynamic, rather than a deterministic, set-up. As the seasons change, or we fly to another timezone, or go on night shift, the clock has to be reset. So what is doing the controlling? The answer lies, of course, with the whole structured system of interactions on the inside, interlocking with a structured system on the outside.

In this chapter, I have tried to show that, even at these basic levels of life there is already tremendous complexity of organisation and structure. The structures, and the information they provide, already turn living systems into intelligent beings, as distinct from cue-response machines. Finally, note that, although the systems described here involve adaptations to change, in each case that change can be characterised as having a fairly *constant* underlying structure. Adaptation – or, rather, adaptability – to it can, therefore, be achieved at the physiological level (including the gene regulatory systems nested within it). Later, we will be considering more complex environments where the structure-of-change itself changes, sometimes rapidly and radically, throughout life. Such situations demand other physiological, behavioural and cognitive systems. Before we turn to consider those, let us consider, in detail, relations between organisms and environments.

4
Bodily Intelligence

It may be stretching credulity a little to suggest that cells think! But they do talk, in a sense. They, and a myriad biochemical agents within them, 'talk' to each other. Quite unlike the silent servants of an autocratic executive power we have come to expect, they chatter ceaselessly to organise themselves. And they do so, not as a token of resistance, but because by such means they 'know' best what is needed.

In this way cells do create intelligence of a sort, in their own way, and that's a great stride for mere molecules, operating without any supervisory intervention or external design. Much of what we have learned about the cell over the last few decades defies the traditional conception of a closed 'machine', responding in routine ways to repetitive cues, purely to maintain an internal equilibrium state. Instead, we are finding a system, even at this primordial level of life, of immense complexity, very much 'open' to the structure of the outside world, and quite creative in response. In this chapter, I want to try to provide a glimpse of that complexity to demonstrate how it constitutes a firm foundation for the evolution of additionally complex systems such as developmental and cognitive systems.

Already we have referred to cell signaling in relation to the coordination of internal and external structure in single cells. It all becomes more complicated in multicellular animals. The organisation of single cells into cooperating groups involves a radical change in relations between the environment and the whole behaviour of the individual cell. The cell does not, now, exist independently, but in a changing profusion of signals from other cells, collectively dealing with another profusion of signals from the changing outside world. It is hardly surprising, therefore, that *all* activity within cells is acutely dependent upon signals from other cells. Without those signals all cell activity stops, as if, like a rabbit in car headlights, frozen by confusion.

Signaling in complex environments

The immediate environment of any individual cell is the extracellular matrix and fluids. It is not by any means an empty space, but contains a throng of factors and signals from other cells and tissues. Nor will these be randomly or evenly distributed: rather they will tend to have some underlying pattern or structure. As with bacteria, sensitivity to this pattern is mediated through cell membranes by means of specialised receptors. A ligand, or signaling protein, in the extracellular fluid, produced from some other cell far or near, binds to a specific receptor. The receptor then undergoes shape transformation, which, in turn, initiates (or can inhibit) responses within.[1] This is by no means a routine process, though. The traditional view of cells passively receiving independent environmental cues to which they respond in isolation from other cues is now being transcended.[2] At any one time the cell surface is being bombarded by multiple inputs simultaneously. Dealing with these as if they were independent cues would only result in utter disorder. Instead, organised reponse requires continuous and precise integration of their 'messages'.[3]

Internal signaling assimilates external structure

This context sensitivity of all cell activities is reflected in extensive *internal* signaling networks. Generally, binding of a ligand to its receptor activates biochemical pathways, including some leading to gene transcription, by triggering cascades of other signals. Some responses, such as those to steroid hormones, involve only two or three steps from initial binding to genetic transcription and synthesis of product. More typically, though, the cascades are modulated by vast networks of internal signaling pathways, themselves integrating, and responding to, the multiplicity of signals from outside. So important is this cell talk that the production of relevant signal-receptor proteins is, itself, a major activity of the cell. The systems are so complex, precisely because they need control points where they can listen to so many other cell states or events. This is why diagrams of signaling pathways in research papers and textbooks increasingly resemble an explosion in a spaghetti factory.

These systems have come to light through the discovery over the last twenty years of an extraordinary variety of signaling proteins.

Indeed, it seems likely that such a variety was an absolute prerequisite for the first multicellular organisms to form, and they have been found in single-cell species that occasionally colonise.[4] As components of communication pathways, internal signaling proteins are sensitive to structural information in a number of ways. They may, for example, undergo change of shape as a result of interaction with a signal or other factor, and so alter conduction along a pathway. They may be moved, or assist movement of other components, to other parts of the cell. Or they may, perhaps with the assistance of another enzyme or cofactor, catalyse a chemical reaction. Any of these functions may be conditioned by the levels of other factors through several levels or depths of interaction.[5] 'Although pathways are often conceptualized as distinct entities responding to specific triggers, it is now understood that inter-pathway cross-talk, and other properties of networks, reflect underlying complexities that cannot be explained by the consideration of individual pathways in isolation.'[6]

One aspect of this arrangement is the way that pathways interconnect and coordinate as hierarchically organised sub-networks, or 'modules', often centred around highly connected nodes, or 'hubs'. These have extraordinary robustness against perturbations and afford enormous flexibility in response to novel external changes. There is evidence from computer simulation studies that modularity only evolves in environments that pose structured change: that is, operate under deeper 'rules' than mere cue-response relations or other simple associations.[7] Changing environments, that is, seem to be prerequisite, but not *randomly* changing environments. One major advantage of such modular structure is that novel ways can be found around prior developmental constraints, opening up new gene expressions, and new adaptabilities, and so altering evolutionary trajectories.[8]

Another indication of this contextual sensitivity has been the failure of models of signaling pathways that treat them as isolated functions, operating independently of what is happening elsewhere. Eric Schadt and his colleagues point out how models of signaling systems as simple, linear pathways have been disappointing: drugs administered on that assumption often turn out to be ineffective. Instead, the pathways are 'best modeled as highly modular, fluid systems exhibiting a plasticity that allows them to adapt to a vast array of conditions.'[9] In this way the same signal (or drug dose) can prompt a variety of responses: for example, cell growth and proliferation in one context and differentiation in another.[10] Indeed, many research groups are now looking for better understanding of cell signaling in that context-dependent manner.[11]

A good example of this coordination of internal and external structure is that of Epidermal Growth Factor (EGF), the subject of thousands of research papers since its (Nobel prize-winning) discovery in 1962. EGF is a peptide produced in the brain and circulating in body fluids where it binds to EGF receptors (EGFR) on many cells. This binding initiates a range of internal signaling processes that (ultimately) mediate a vast range of responses such as growth, cell division, differentiation, migration and so on, depending on physiological context. The interactions, involving dozens of different internal components, often mean that the same receptors on different cells can activate quite different internal signaling pathways. Specific activation is conditioned by the extracellular context, from the composition of the extracellular matrix to the finer, spatiotemporal composition of surrounding fluids. But much more needs to be discovered about this. As Stanislav Shvartsman and his colleagues say, a key challenge is to integrate what is known about internal signaling mechanisms with the spatiotemporal dynamics at various levels outside the cell, including the tissues and the whole organism.[12]

Another example is the G-protein-coupled-receptor (GPCR).[13] This ubiquitous set of receptors (over a thousand varieties have been identified in different tissues) mediates most of our physiological responses to hormones, neurotransmitters and environmental stimulants. When you see light, taste something nice, have an adrenaline rush or use beta blockers, you are sending signals to your GPC receptors. Like other receptors, they function to initiate or suppress a multitude of biological processes within the cells. They play a critical role in heart disease, blood pressure regulation, inflammation, psychological and many other disorders. Not surprisingly, then, more than half of all drugs given to patients work by targeting one or another GPCR on the body cells. Again, GPCR stimulation takes place against a background of structured activities, so recent research is exploring the ways in which signals are integrated synergistically on the receptor and/or in crosstalk between receptor proteins.[14]

Such structured background has also been studied in a group of signaling agents called chemokines. These include a variety of substances secreted by cells to induce motion in other cells elsewhere in the body. For example, they might induce immune cells to move quickly to a site of infection; or other cells to move to a site of tissue repair; or cell migration during normal development. So far, over fifty chemokines and twenty chemokine receptors have been identified. But, in each case the recipient cell must gauge the spatiotemporal disposition of the signal through its specialised surface receptors, as described for bacteria in the previous chapter.

In an article, the title of which begins 'Orchestrating the orchestrators', Shannon K. Bromley and colleagues have reviewed the complex temporal and spatial patterns of expression of chemokines and their interactions on receptors.[15] In another study, Kentner L Singleton and colleagues studied the interaction among thirty 'signaling sensors', ranging from membrane receptors to gene transcription factors, and found that 'spatiotemporal patterning controls signaling interactions...in a physiologically important and discriminating manner'.[16]

Knowledge of such interactions is important in many practical spheres. The bacterium *Staphylococcus aureus*, now famous for its role in hospital acquired infections, creates and extends infection partly by blocking the normal binding of chemokine to its receptor on immune (white) cells. The bacterium produces a rival protein that binds to part of the receptor, such as to change its 3D shape. This inhibits the effect of the chemokine, and transmission of signal to the inside of the cell. The result is to freeze the normal motion of the immune cells, keeping them away from the point of infection, and leaving the bacteria free to multiply and do their work.[17]

The structured matrix

Most of the extracellular environment consists of this kind of biochemical 'soup' with signals hitting the cell surface something like a meteor shower. But cells are also physically attached to the extracellular matrix – the underlying connective tissue – and to neighbouring cells. Many changes occur in the physical properties and topography of this three-dimensional environment, including mechanical push–pull forces, fluid turbulence (as in blood vessels), compression from other cells and so on. Cells sense these changes through additional receptor complexes on the cell surface: changes that have been shown to influence the metabolism, growth and organisation of cells in tissues (cancer, indeed, is sometimes thought of as a disease in which the physical regulation between cells has malfunctioned).[18]

By their nature, these physical forces are exerted on the cell surface in structured spatiotemporal forms. The cell responds to the mechanical forces and deformations by translating them across the cell membrane into spatiotemporally organised biochemical signals. These are transmitted to internal signaling pathways – or, more rapidly, through filaments of a cytoskeleton – with widespread internal effects, including change of shape, cell division, cell death and so on.[19] Again, the importance of this environmental sensing system to the cell is seen in its formidable complexity. Benjamin Geiger and colleagues estimate that just one of the receptor-signaling complexes involves around 160 molecular components. The entire network of that one complex comprises around 700 links, about half of which are interactions in which activity of one component affects that in others.[20] As with the other networks, mentioned above, they have all evolved to bring the internal structure of the cell into closer interplay with that of its structured environment.

How genes are used

Many signaling components serve as transcription promoters or inhibitors, affecting the use of genes in the cell through long cascades of regulation. Proteins transcribed from some genes may very well enhance or repress the utilisation of other genes. And this may affect still others, and so on, in 'downstream' cascades of regulation, so that the final product (for development or for cell metabolism) results from vast networks of interactions. Indeed, it is now known that up to 95% of genes are regulatory in that sense, only the small minority transcribing as structural proteins that are actually used in cell maintenance or development. These gene interaction networks, many of which have been conserved across evolution, from yeast to humans, show considerable flexibility in balancing specific needs with wider context and longer term demands. For example, the so-called rapamycin (mTOR) signaling pathway, in mammals, senses, and responds to, nutrient availability, energy sufficiency, stress, hormones and other factors to modulate protein synthesis.[21]

In addition, the actual products of gene transcription can be modified or otherwise regulated in a vast variety of ways, according to the current needs of the cell, the organism or its current state of development. The actual transcription of genes takes place first into an intermediary messenger RNA (or mRNA). This serves, through other intermediaries, as a template for the assembly of a string of amino acids to form peptides and proteins. It has been known for some time that the components of the mRNA can be rearranged *after* transcription to form a variety of *alternative* templates.[22] This process can have a variety of effects according to what is optimal for the moment. One obvious result is that different proteins can be produced from the same gene, with potentially widely different functions. In addition, it is known that modified mRNAs have functions other than serving as protein templates. For example, they can increase efficiency of initial transcription of the gene; improve transport of products from the nucleus; improve translation into proteins; and increase rate of decay of the mRNA product itself.[23]

The picture of how genes are used, now emerging, is, of course, a far cry from the traditional deterministic model of one gene producing one protein, and that's all!. But the importance of these processes is indicated by the fact that at least 74% of human genes are 'alternatively spliced' in this way.[24] Moreover, across species, *numbers* of possible alternatives correlate strongly with levels of evolutionary complexity.[25] Finally, this kind of shuffling can explain the rapid evolution of novel genes, and of new species as in the 'Cambrian explosion' of around 500mya.[26]

But there have been other revelations. Other significant gene regulatory pathways have come to light recently in the form of so-called noncoding RNA genes – genes that yield RNA but do not produce proteins *at all*. These are not marginal items of curiosity: they have been found to utterly dominate gene production, accounting for up to 98% of it! Interestingly, ncRNAs seem to have become more important in more complex organisms, where they seem to have a greater role in developmental control of gene transcription than regulatory proteins. Some have been shown to influence the kinds of regulation described above, as well as the developmental timing of protein expression.[27] Others seem to interact with hormone receptors, modulating promoters, silencing genes or acting as coactivators of transcription. Little wonder that the discoveries of these new functions of RNA were hailed as 'Breakthrough of the Year' in 2002 by the journal *Science*.

Other forms of post-transcriptional protein modification are being discovered all the time, each with stunning new implications for cell complexity.[28] In all these ways, genetically identical cells can come up with radically different products. This has, of course, encouraged new views of the cell, and of the nature of genes, quite incompatible with simple deterministic models and rigid programs. Instead of a pre-written song sheet, each cell seems to sing to any one of a myriad possible tunes constructed in context.

These recent findings explain why the very concept of the gene has undergone considerable revision in recent years. It is no longer valid to speak of the gene in the traditional metaphors of recipe, controller, designer and so on. Rather we now see genes as partners in vast networks of regulation, being used *by the system* for the synthesis of important components in an interactive cell dynamics. In those dynamics, control is distributed in what Y. Bar-Yam and colleagues call a 'democratic' way among many players.[29] Whereas we usually think of body, mind and their functions being created by the genome, in important senses the genome is created by those functions and the structure of environmental changes around them. These regulatory dynamics probably explain why there is little association between numbers of genes and species complexity. Vertebrates, on average, have only about twice as many genes as invertebrates, in spite of vast differences in complexity: the simple nematode worms have nearly 20000 genes; fruitflies around 14000, and humans around 25000 (and, oh yes, carrots have around 45000!).[30]

Again, the increasing dominance of dynamics over elements is precisely what is required by the kind of complex environments cells and organisms actually experience. The evolutionary implications of this reality explain, of course, the gathering demands for the 'extended evolutionary synthesis', mentioned in Chapter 1.

Cell dynamics

The non-linear, interactive nature of these networks is also being discovered through new conceptual and analytical tools. The intense crosstalk between signaling pathways within cells provides for complex non-linear responses, creating 'criticality' in cells – a threshold across which a system can transform from old to new behaviour. These dynamics permit creative, adaptable responses to the outside world, rather than slavish, and ultimately fruitless, stimulus-response mechanics. Such criticality has been well demonstrated in cells.[31]

Generally, in this view, the system forms a global state space within which the cell itself is a basin of attraction. But so are the myriad substates within the gene regulatory and signaling networks.[32] The dynamical properties of the networks allow the system to jump smartly among a variety of stable attractors appropriate to different input states or perturbations, including the ability for novel solutions. The networks can also undergo phase transitions into still more integrated networks (a set of smaller networks becomes a bigger one). These, in turn may exhibit emergent properties such as wider integration of signals across space and time, new feedback loops, and new levels of regulation, all permitting the creation of novel outputs and adaptabilities.[33]

The same picture applies more specifically to the utilisation of genes. Through self-organising dynamics, some transcription factors of genes can have their interactions modified by various other factors in the cell, in effect 'rewiring' the gene network.[34] For example, a deficiency in the provision of a metabolite, either from the environment, or through a genetic mutation, can be overcome by recruitment of an alternative biochemical pathway. As Andrew Wuensche puts it, 'Each subnetwork settles into one of a range of attractors according to its current state, which if perturbed can cause the dynamics to jump to alternative attractors.'[35] Crucially, such dynamics also permit the creation of new expressions that can become new targets of natural selection, in turn modifying trajectories of future evolution.

This whole arrangement of dynamically structured signaling networks has evidently been remarkably successful. It is possible to represent nearly all of those known in all existing species on a single chart, reflecting a striking underlying uniformity within the great diversity of biological systems on earth.[36]

Physiology

The coordination of cell–cell signaling systems across the body as a whole, in the context of those outside changes, is what is called physiology. Dealing with those changes are very much lively, intelligent systems, as we have seen. Unfortunately, the view of physiology as a set of systems for merely maintaining internal equilibrium has persisted through the popular model of homeostasis. In this model, each aspect of physiology has been viewed as independently preserving, as far as possible, some constancy of part of the internal milieu – blood sugar, temperature, fluid balance or whatever – in the face of disturbances from inside or outside. As Steven Rose points out, 'No modern textbook account of physiological or psychological mechanisms fails to locate itself within this homeostatic metaphor.'[37]

However, the metaphor is somewhat misleading. Like other regulatory processes, physiology is not a question of recruiting If-Then homeostatic mechanisms, as if operating a simple equilibrium machine: physiology is not just a set of internal programs responding to external triggers or cues. It has been known for a long time, for example, that regulation and integration of organic activities takes place on various hierarchical levels.[38] Non-linear dynamic (even chaotic) models may be more accurate in reflecting the often variable, adaptable responses in physiology: very flexible and easily altered, yet still regulated.[39]

Take, for example, aspects of the endocrine system: many hormones are associated with the brain's sensitivity to changes in the outside world, as part of wider neuroendocrine systems. Hormones such as thyroxin, steroids like the sex hormones, and glucocorticoids, which regulate cell/tissue metabolism, are closely involved in the daily integration of body systems. As with other intercellular signaling, this integration is mediated through hormone receptors on cell membranes to which the hormones bind. The integration happens because hormones tend to meet, at the cellular level, with *interactive*, rather than independent, effects, reflecting the structure of environmental events. That is, responses of hormone receptors are cross-regulated, depending on the hormone – and wider bodily – context.

An example of this physiological sensitivity to environmental structure is the regulation of metamorphosis in tadpoles. Circulating thyroid hormone tends to initiate metamorphosis in tadpoles by binding with cell receptors, initiating gene expression and developmental pathways. However, the effect seems to depend on interaction with corticosteroid hormones, circulating levels of which increase dramatically in response to environmental stress such as pond drying.[40] So the system's monitoring of environmental stress modulates the effects of thyroid hormone, in turn controlling the timing and coordination of metamorphosis. In fact, it is now well known that this neuroendocrine stress axis is a common mechanism for monitoring the structure of environmental factors in vertebrates.

Another example of this context-sensitive interplay concerns regulated potassium (K+) secretion by cells of the kidney into the kidney tubules (and subsequent excretion). This regulation is critical, because high concentrations of potassium in the body can cause cardiac arrhythmias, or even cardiac arrest. But the rate of secretion does not depend only on plasma concentration of K+. That simple relationship is conditioned by sodium concentration in the kidney tubules, circulating levels of the anti-diuretic hormone (ADH), the hormone aldosterone (which itself is influenced by other factors, such as stress) and other factors, all of which are influenced by other bodily states. Although such processes may look superficially like simple, independent, homeostatic systems, based on threshold switches, they are actually emergent properties of interactions of many parameters both within and between systems. This synergistic interplay in physiology is better described as homeodynamics.[41]

The neuroendocrine stress axis – or to give it it's full name, the hypothalamic–pituitary–adrenal axis (HPA or HTPA axis) – is a key physiological system. It regulates responses to stress, either from internal or external sources, and also affects many body functions such as digestion, the immune system, energy metabolism and emotional aspects of psychology (i.e., feelings). Part of the classic stress response consists of secretion of corticotropin-releasing hormone from the hypothalamus in the mid-brain. This hormone passes quickly to the nearby pituitary gland that actually releases the hormone corticotropin. This then enters the bloodstream to reach the adrenal glands above the kidneys. In the adrenal glands it induces release of cortisol into the bloodstream, followed by aspects of the classic 'stress response' such as feelings of alarm, muscle tension and increased heart rate.

However, there are many other players in the stress-response system. One is noradrenalin, produced in the *locus coeruleus*, a small nucleus in the hindbrain. The noradrenaline is released via nerve fibres into numerous parts of the brain, in response to perceived challenge. It promotes a state of excitement and awareness (as well as release of corticotropin-releasing hormone from the hypothalamus, as just mentioned). The other player is adrenalin, which is released from the adrenal cortex following stimulation from nerve branches terminating there, themselves being stimulated by fibres from the *locus coeruleus*. Together the hormones produce the classic preparation for 'fight or flight' including increased heart and respiratory rates, dilation of arteries to muscles (with constriction of peripheral blood flow), release of blood sugar for energy, increase in blood pressure (to get blood to the muscles) and suppression of the immune system.

Although superficially plausible, the general 'stress-response' concept has turned out to be too simplistic. Responses now appear to be more variable, depending on individual histories and current contexts. Also, chronic stress (real or perceived) can have complex psychological and neurological consequences in animals and humans, including emotional dysregulation, panic attacks, posttraumatic stress disorder and many other states. On the other hand, the normal working of the systems, together, as they were evolved to do, 'results in the stress instruments producing an orchestrated "symphony" that enables fine-tuned responses to diverse challenges'.[42]

Given the pervasiveness of this system in physiology, with its numerous interfaces and feedback processes, it can readily be modelled with non-linear dynamics. Instead of independent cue-response functions, NLD models would propose the emergence of new levels of regulation, consisting of basins of attraction, from the interaction of subsystems. The state of the whole system will tend to be drawn towards one or other of those attractors under challenging conditions. This explains how it is not the simple presence or absence of a stress situation that determines the response, but the sense of control over it – a consideration that has become increasingly important in human societies.

Indeed, non-linear dynamical models are being increasingly used as analytical and conceptual tools in studies of physiology.[43] Remember that the main, general advantage of non-linear dynamical, over deterministic, systems is the rapidity and creativity of response to perturbations. The more distinctive properties of NLD in physiology, however, include the following: (i) they don't tend to settle towards one specific (equilibrium) state, but exhibit numerous possible states in readiness for changing conditions; (ii) activities arise as much from internal feedback as from external conditions, with predictability derived from the much deeper structure among multiple variables; (iii) disease, or other malfunctioning, reflects a breakdown of the deeper structure of the system – that is, reduction of chaos. The self-organising properties of non-linear dynamics, that is, brings a deeper structure and order to physiology.

One example of NDL research in physiology is that on cardiac functions by Ary Goldberger. Contrary to the predictions of a homeostatic model, the behaviour of the healthy human heartbeat shows deeper statistical structure than surface monitoring of the heartbeat might suggest, even under resting conditions. These properties suggest that 'non-linear regulatory systems are operating far from equilibrium, and that maintaining constancy is not the goal of physiologic control.'[44] This means that the variation in heartbeats has deeper, more contingent, structure than previously suspected: indeed, that chaotic dynamics in the heart are found, not in disease, but in the normal sinus rhythm.

Remember that the deeper structure in such dynamics makes them robust and more resistant to injury or deviation. Disease states, such as narrowing of arteries and veins, high blood pressure and the ageing process, reduce the adaptability of cardiac dynamics. Goldberger says that most physiological systems are like that. He suggests that many other measures of biological systems that take account of multiple spatial and temporal variables reveal considerable correlational structure.[45]

This kind of complex variability, rather than a regular homeostatic steady-state, appears to define the everyday function of many biological systems as well as physiological processes. As Hector Sabelli and Aushra Abouzeid note, variation in heart rate reflects the *totality* of our physical, mental and emotional state as we interact with changes around us.[46] More structured regulatory processes are adaptive because they serve as an 'organising principle' for non-linear, non-equilibrium (i.e., constantly novel) changes in the physical and social environment. Under such real-life conditions, 'mode locking' in a single steady-state would restrict the functional responsiveness of the organism. As Goldberger and colleagues suggest, 'A defining feature of healthy function is adaptability, the capacity to respond to unpredictable stimuli and stresses', whereas, 'highly periodic behaviors...would greatly narrow functional responsiveness.'[47] It is the breakdown in such deeper, integrative responsiveness in changing conditions that produces disease states.

Most of the physiological systems' interface with the outside world is through the nervous system, but especially through a tiny nerve centre in the brain called the hypothalamus. The hypothalamus receives nerve impulses from all sensory systems (vision, hearing, smell and so on) themselves responding to changes in the outside world, as interpreted through 'higher' centres in the brain. It also has rich beds of chemoreceptors monitoring the internal state of the body. It responds by issuing certain signals to endocrine glands, as well as creating many aspects of feelings, (which we will examine in a later chapter). Much of this is done through the neighbouring pineal gland that, in turn, secretes a wide range of hormones into the circulation to influence functions elsewhere in the body. So interconnected are the neural, endocrine (hormonal) and immune systems, in fact, that they are often referred to as a single (NLD) system having interactive roles in almost all body functions and diseases.

I hope to have provided just a glimpse at the complexity of intelligence that exists within the communications among signalling systems in cells and bodies. I have referred to a wide range of species from yeasts through flies and frogs to rats and humans; but, really, so much of this intelligence has already evolved in simple creatures and even single-cell organisms such as bacteria. I hope it clearly demonstrates that rich foundations for more complex traits and functions emerged from natural, self-organising processes at an early level in evolution. These became readily extended in developmental, neurological and cognitive systems to reach much greater adaptabilities, as I hope to show in the next few chapters.

5
Evolution of Development

The apparent mystery of evolution, in which original ensembles of molecules come to have the structure and complexity of fully living, intelligent forms, seems to be repeated in compressed form in the development of individuals. The changes that take place in living systems from their moment of conception – and, even more, the intricate patterns that they form – have intrigued observers since Ancient Greece. How can complex structures arise from simpler ones, along an orderly trajectory, in such a short period of time? This is the apparent mystery of development. In this chapter, I want to show that development is being slowly demystified: but, more importantly, that it is not merely a passive process of growth to maturity from smaller origins. Rather development has itself evolved as another crucial bridge to the evolution of complex intelligent systems: in fact, in each organism, development *is* an active, intelligent system in its own right.

The process seems complex enough even from observation of the visible features of bodies and main organs. But a closer look through a microscope makes the process and its results seem quite stupendous, especially in the more evolved, complex organs such as the brain. Nicholas C. Spitzer points out, somewhat poetically that, in the brain, 'Specifying how nerve cells differentiate is a Herculean challenge: a huge number of neurons with extraordinary diversity of elaborate architectures and sophisticated functions materialize almost magically during a brief period of embryonic development.'[1] And that is not even to mention the emergence, from that, of even more complex functions of cognition and, in humans at least, all else that make up the mind.

In trying to explain that 'magic', most non-scientists, and many scientists, have clung to a tacit teleology prevalent since Aristotle: that is, of some inner 'purpose' built in to the fertilised egg. Along with that has been the view that such complexity can only arise from a clear 'plan' or 'recipe' in the genes inherited from parents, more or less attenuated by a 'good' or 'bad' environment. In that view, development consists of an 'assembly and growth' process, or maturation of something already 'there'. It starts with a definite plan in the genes, and ceases at a definite end point. The mature animal then has all the form and abilities needed to cope with the real world, independently of parents or other adults. So developmentalists such as Massimo Pigliucci complain how neo-Darwinism, and the tendency to view the mature organism (the 'phenotype') as the direct product of the genes, essentially leaves more important adaptive roles for embryology and development out of the picture.[2]

Only in the last two decades has the study of development become, once again, integrated within an ecological framework of environmental change and structure. The alternative view of development now regaining popularity, though actually around for decades, is that of a constructive, adaptive process based on the interactive dynamics within cells and with local circumstances. We are learning how, instead of a passive process of adaptation to stable or recurrent circumstances, development is a far more active process of coping with change. More importantly, we are learning that, far from being merely a means to an end – the maturation of a juvenile from birth to adulthood – in the long run, development has become an end in itself in the management of lifelong adaptability. The realisation provides a crucial conceptual bridge in the understanding of evolution in general and of complex systems in particular.

As a result, development of organisms from the single cell or fertilised egg is now a huge area of investigation, and I make no attempt to do justice to it in a short chapter. Instead I want to make a more special point. We have seen how the self-organising systems in living cells and bodies continuously adapt to changing environments by taking account of its deeper structure. Here, I want to show that, as species have evolved into more complex environments, development itself has become an increasingly important means for dealing with them.

Significantly, a new focus on this more 'active' development has integrated evolution and development more closely, and indicated, not just how development has itself evolved, but also how evolution has been influenced and changed by it. This alliance is fundamentally amending the traditional evolutionary picture and helping us to understand the emergence of more complex living systems. As Gerd Müller says, the gene-centred 'Modern Synthesis', which has reigned over the last half century, glossed over the problems of innovation and complexity by assuming that genes, acting in linear fashion, are the sole causes of variation in structure.[3] 'Evo-devo' (as it's now called) offers a new way of viewing evolution of complexity. The upshot is that organisms are both more 'evolvable' and more 'developmentable' than we ever previously imagined. This insight is proving to be an important link to understanding the evolution of complex cognitive systems.

Becoming different

The creation of a juvenile from an adult is relatively easy in single-cell organisms such as bacteria: they just copy critical components, including genes, and then split into two identical cells by binary fission. In multicellular organisms, with many different kinds of cells, a more complicated process is obviously called for.[4] In some cases the parent just fragments, and juveniles develop and grow from the pieces. In others, reproduction is performed by 'budding' from the parent. Small clumps of cells pinch together, and then detach to form new, immature individuals. In more evolved animals, with sexual reproduction, development starts through combination of separately produced eggs and sperm. The fertilised egg then undergoes several stages of division to form an embryo. The subsequent production of new cells and tissues, with all their materials in the right place, in the right proportions and at the right time needs to be a highly coordinated set of processes. Moreover, that distribution needs to be integrated with what is happening in the rest of the embryonic or juvenile individual.

Actually, it's not just a problem of putting cells in the right place: but of putting *different* cells in the right place at the right time. After all, the whole point of multicellularity is division of labour among more specialised groups. It needs to be remembered that all cells are originally undifferentiated, all carrying the same set of genes and a host of other components. We have been reminded of this in recent debates about the use of stem cells for tissue replacement or repair in humans. So how does that differentiation and disposal – with clear borders between groups – come about?

There have been various proposals, all presenting further examples of cells 'talking' to each other. In the 1950s, mathematician Alan Turing proposed a concept of 'morphogenesis'. This is based on the diffusion of chemicals between cells, regulating how they react and develop in a concentration-dependent manner.[5] In the 1960s, Lewis

Wolpert stressed how morphogenetic gradients gave cells positional information whereby they 'know' where they are within the mass of other cells in a developing organ. The idea has been well supported as we shall see.[6]

Other models have been based on more direct signaling processes. For example, cells may express a ligand that binds to receptors on neighbouring cells. Receptors then act, through signalling cascades, as transcription regulators on genes, with appropriate component production. In some cases, at least, the recipient cells can send reciprocal signals to the source of the ligands to stimulate or inhibit more ligand production, so influencing the extent to which neighbours become the same or different.[7] In each case, the effect of the signal is that some genes will be recruited, while others remain unused, different products result and cell development takes different pathways.

One obvious consequence of those signals is that development of cells and tissues is highly context-dependent. Indeed, *Science* magazine's 'breakthrough of the year' for 2008 was the experimental 'reverse development' of skin cells back to stem cells which then, in a different context, could be grown into completely different cells.[8] Early brain development in vertebrates presents a good example of this context-dependency. The fertilised egg rapidly divides into a ball of cells that then begins to differentiate. From these, one distinct variety forms a ribbon of cells called the 'neural plate' on the dorsal (back) surface. A groove in the neural plate deepens, and the sides curl around to form a 'neural crest' on either side of the neural plate. The sides of the crest eventually meet to form the neural tube surrounded by the ectoderm, the outer layer of cells. Interactions between the ectoderm and the neural plate are thought to start the differentiation of the neural crest cells into (ultimately) many different kinds of brain and spinal cord tissues. The interactions involve multiple signals, in precise spatiotemporal order.[9]

Within the cells, meanwhile, waves of signaling and gene regulatory processes begin. These recruit genes in an ordered sequence, so products become assembled in self-organised patterns. Those patterns are themselves varied according to the kinds and structure of their external contexts.[10] This was illustrated in an experiment as follows: 'We exchanged neural crest cells destined to participate in beak morphogenesis between two anatomically distinct species. Quail neural crest cells produced quail beaks in duck hosts and duck neural crest produced duck bills in quail hosts. These transformations involved morphological changes to non-neural crest host beak tissues. To achieve these changes, donor neural crest cells executed autonomous molecular programs and regulated gene expression in adjacent host tissues.'[11]

This sensitivity to context, though, is not just to local 'triggers', or cues, but also to the environment of development as a whole. This has been demonstrated in the way that the size of a body part is not absolutely determined locally: it also depends on the sizes of parts all around the body. For example, butterfly larvae whose hind wing discs (primordial wings) were removed, developed into adults with forewings that were disproportionately large for their body size. Additional changes were found in the relative sizes of the thorax and forelegs but not in the head or abdomen.[12] It's as if the system has an intelligent awareness that it needs to compensate for the absence of rear wings. Such relationships are known as *allometric*, and Charles Stevens has shown how they reflect the role of 'power laws', or sensitivity to deeper structure, in the process of development.[13]

An example: development of body axis

Cell growth and differentiation from the egg proceeds to form a ball of cells (the blastocyte stage), the establishment of initial anterior–posterior (front and back) polarity, and the formation of three primary layers (gastrulation). From these layers are formed all the tissues in the embryo, and the basic body pattern of skeleton, muscles and organs. Research over the last twenty years has described the signaling pathways, and the developmental gene regulations, through which cells become sharply committed to one lineage or another.[14]

The first developmental task in the formation of the embryo is the creation of necessary polarities, front to back and top to bottom (antero–posterior; ventral–dorsal). This must be followed by the differentiation of body segments, and of important structures such as head, tail and internal organs. Most of the research has been done on fruitflies, which have a conveniently short breeding cycle, are readily available and easily kept in laboratories. But the results are remarkably similar across the animal kingdom. In no case does a 'groundplan', blueprint or fixed program seem to operate. Rather, body form emerges dynamically from the interactions among hierarchies of regulatory factors within and without the egg and developing embryo. The molecular geneticist Enrico Coen has likened the whole process to an orchestra without a conductor.[15]

Perhaps the most-studied example of self-regulated development concerns a subset of the *Homeobox,* or *Hox,* family of genes. The protein products of *Hox* genes serve as transcription factors – promoting or inhibiting transcription from the genes – in a wide variety of developmental pathways. Their involvement in early development is universal across species from flies to humans: an amazing commonality underlying all the diversity of animal types. Clues as to their function came from laboratory-induced mutations that produced weird effects of body parts in the wrong place, such as a leg growing where an antenna should be. Such mis-expression, plus the fact that very similar *Hox* genes are found in a wide range of species, of very different morphologies, suggested that they act as developmental regulators for positioning and timing of development, rather than structural codes.

One of the striking aspects of the *Hox* family of genes, indeed, is the way that they are utilised in a precise spatiotemporal order. Corresponding with this is their positioning on the DNA strand in an order roughly in parallel with their respective targets of influence from head to tail. The sequence of recruitment of them in early development follows that order to obtain the precise antero–posterior differentiation of rapidly dividing cells. Indeed, it seems that this recurring need for a spatiotemporal order of expression is what has kept the cluster together over 600 million years of evolution.[16]

It is now known, however, that the effect of Hox proteins on gene transcription, and associated developmental pathways, is conditioned by various cofactors. This ensures that they work as required in the right places at the right times. Hox proteins can not only act as transcription activators or repressors, depending on where and when they are distributed, but also regulate the expression of numerous *other* regulatory genes (and other cascades of signals), also in context-dependent fashion.[17] But they are, themselves, regulated in order to achieve a highly coordinated process. While the Hox protein diffuses along the antero–posterior axis in one cascade, it is being met by the cascade of another regulator, diffusing along the dorso–ventral axis. The timing of the respective flows of these factors – that is, their meeting in a precise order – establishes the so-called Hox clock, turning temporal order into spatial disposition.[18]

In fact, it is now known that much axis formation is induced in the oocyte, or future egg, even before it's laid by the mother. This occurs through products of the *maternal* genes being deposited in the egg beforehand. These, too, have regulatory functions operating through networks of associates.[19] For example, the maternal messenger RNA for a protein known as 'Bicoid' is deposited in the egg, but it is unevenly distributed by the products of two other maternal genes to form a gradient from front to back. This Bicoid gradient determines where and when

Hox transcripts enhance or suppress structural genes, thus influencing eventual head and tail differentiation.[20] While Bicoid is determining the anterior section of the larva's structure, a different morphogenetic gradient is being set up at the posterior end. This consists of the mRNA Nanos. It influences tail formation in the fly but critically depends on the presence of at least two other proteins. A similar choreography presets the embryo for the effects of Hox regulatory transcripts in establishing the dorsal–ventral axis.

All this is an example, of what is called 'combinatorial regulation'. Through such regulation, different proteins can associate in different ways, with different consequences for transcription, according to local and wider context. It should be noted that what I have just offered, though, is a mere glimpse of these processes. Much is still to be worked out and there is much debate about interpretation even of current findings.[21] However, although most of the above description is from research on fruitflies, similar processes have been found to operate in other species.[22]

The point is that the production of distinct types of cells (rather than indistinct gradients), in just the right place at the right time, reflects a highly dynamic process dependent upon the structure of an internal network of influences. Rather than a system of command genes issuing instructions for development, these combinatorial interactions suggest, as with most other aspects of development, 'the dynamics of a cocktail party where the composition and themes of individual conversation groups change with the stepwise exchange of participants.'[23] For such reasons cell fates have themselves been described as high-dimensional attractor states in such dynamic gene regulatory systems.[24]

What we know so far, then, implies the resolution of developmental trajectories, in very flexible ways, in response to the dynamic interactions among a host of factors. Of course, it might have been easier for natural selection to have produced a deterministic command system for development. But that would have presupposed fairly stable environmental conditions during development. A dynamic system of regulation – called 'epigenetics' (or 'around the genes') – is more able to deal with changing environments, with robust, creative responses. These may either conspire to maintain development on a particular trajectory or produce more variable development, depending on the structure of the external environment. Let me now try to illustrate a little more of the relationship between developmental constancy/plasticity and the structural characteristics of the environment.

Epigenetics 1: Canalised development

Many environmental conditions are sufficiently constant from generation to generation to allow the development of fairly uniform structures in organisms, with little variation between individuals. This is the classical Darwinian natural selection scenario, and the beaks of Darwin's finches are good examples, along with numerous other aspects of all living things. Development of those structures also seems to follow a uniform course. However, assembling such structures from a myriad ingredients is still a delicate business, which many environmental fluctuations, bumps and shocks can potentially disrupt. Within the individual organism, too, harmful mutations to genes required for development can disrupt the provision of materials and regulators also needed for development. So how do we get such uniformity? The answer lies in regulatory processes that have evolved to protect development against such disruptions in critical bodily structures and functions. This process is known as the 'canalisation' of development: environmental canalisation protecting against perturbations from outside; genetic canalisation compensating for genetic deviations.[25]

Canalisation of development was first studied in the 1930s and 1940s, largely by Conrad Waddington. He claimed that the development of the basic body parts of organisms is so strongly buffered against genetic variation or environmental perturbation as to be maintained on its 'chreod', or fated pathway. Canalised development is seen in one sense in the differentiation of tissues into distinct types (rather than one type 'shading' into another), which is what lead Waddington to the concept. Again, however, it turns out that the process stems from a self-organising regulatory system, involving signalling and gene regulatory pathways, as well as special genetic mechanisms such as 'back-up' genes.[26] So interactive and mutually compensatory are those regulations that, in experiments, it has been found necessary to disrupt several of them at once to produce variation in the development of critical morphological characters.[27] In many cases the total suppression of a gene, normally involved in development of an organ or tissue, has no discernable effect on the end result of development in the phenotype.

Canalisation of development has been well illustrated in the eye in the fruitfly. This intricate organ, crucial for fly survival, consists of 800 little facets, each facet consisting of 8 light receptors and various other cells, all put together in a very precise pattern. Geneticists have identified the dozens of genes associated with eye development and report a great deal of variation in them from individual to individual. Yet the eyes that develop are virtually invariant from individual to individual, and within a wide range of environmental conditions.[28] The possible consequences of variable genes and environments appear to have been transcended by the dynamic developmental system. And this is typical of numerous other characteristics studied.

Almost by definition, the degree to which a feature has been canalised is not obvious on the surface, so that much genetic and environmental variation remains hidden. An example studied much lately concerns the effects of 'heat shock' proteins in fruitflies. As their name suggests, this is a collection of proteins that seem to offer protection against sudden temperature changes during development. The ubiquity of such protection in flies is such as to suggest little underlying genetic variation in most traits. But, when transcription of heat shock proteins is chemically inhibited, huge variation in nearly every structure, previously showing little or no variation, appears.[29] The effect is rather like a population of humans suddenly producing a generation of offspring exhibiting a tenfold increase in variation in things such as height, facial appearance and developmental defects. It indicates how much of the genetic variation in organisms is normally buffered in development by regulatory interactions.

The details and mechanisms of such regulation are still not well understood. However, recent work by molecular biologists and developmental biologists has brought fresh insights. The mechanism may reside at one or several different levels such as transcriptional regulation, or the kind of network modularity mentioned in Chapter 4. For example, 'input' factors, such as levels of the Bicoid protein mentioned earlier, can actually be quite variable. Interactions between the gene products reduce the effects of such variance, inducing convergence of the developmental pathway. Computer modelling suggests that this is a non-linear dynamic process, resulting in the formation of only one of several possible attractor states.[30]

A most important point to note, here, is that canalisation has evolved for developmental adaptation to aspects of environment that, like seeds for finches, reliably recur across generations.[31] This will hopefully put into perspective the use of the term as a kind of metaphor for development of all traits and functions, including mental and cultural traits. To some theorists, as well as the popular mind, canalisation is often confused with the notion of predeterminism in development. For example, Cathy Dent-Read and Patricia Zukow-Goldring suggest that 'canalization determines the developmental pathways for all members of a species', and at 'all levels'.[32] Howard Gardner has stated that: 'The plan for ultimate growth is certainly there in the genome; and ... development is likely to proceed along well-canalized pathways'.[33]

Such misconceptions are also reflected in the use of expressions such as 'it's genetic', 'innate', 'genetically determined', 'in the genes' and 'hard-wired'. Their use reflects serious misunderstanding of the very dynamical nature of development, and its relationship with evolution. As we saw in Chapter 4, all that is in the genes are sequences of nucleotides that correspond (more or less) with the sequences of amino acids in useful proteins. All that develops emerges from the way that the sequences of nucleotides are utilised by an elaborate network of factors in a non-linear, self-organising system. All appearances of replicable development, and statements about genetic constraints, genetic programs and so on, are really only reflecting the way development has been canalised. As Enrico Coen reminds us, there is no plan for development, at all; development is inseparable from its execution.[34]

Epigenetics 2: Developmental plasticity

These points are demonstrated perhaps more strongly in another set of developmental processes. Many aspects of the environment remain relatively constant across generations, making canalisation an appropriate developmental strategy. But other aspects change significantly, both within and between generations. If there is significant change between generations, then selection for stable endpoints cannot be sustained: canalised development towards a singular, uniform phenotype, within and across generations, may result in a misfit. For adaptation in rapidly changing environments – different, at least, from that of parents – other types of traits and systems have evolved that permit degrees of developmental plasticity between generations. In these, as Clifford Stearns explained: 'It makes sense to leave the choice of ... morphology, and with it an entire trajectory of growth and reproduction, until after the adult habitat has been reached'.[35]

Almost all animals and plants exhibit some degrees of developmental plasticity in at least some characters. A popular example is that of the water flea, Daphnia. If the juveniles develop in the same body of water as a predatory midge larvae (*Chaoborus*), they develop a protective 'neck spine' or 'helmet'. These defensive structures allow the Daphnia to escape from their predators more effectively. This structure is completely absent in their parents that have developed in predator-free water. The induction appears to be due to some chemical hormones (kairomones) inadvertently released by the predator. Moreover the effect is transgenerational, as if inheritance is non-genetic. Those Daphnia born with large helmets tend to have *offspring* that develop helmets, even if they develop in water devoid of predators.[36] Numerous other cases of predator-induced plasticity have been studied. Tadpoles of the wood frog (*Rana sylvatica*) that grow in water previously containing predatory dragonfly larvae – so that the water contains some chemical produced by the larvae – develop bigger tails that allow faster swimming and turning.[37] A species of barnacle reacts to the presence of predatory snails in its environment by developing a 'bent' form that is more resistant to predation compared with the more typical flat form.[38] Other cases of developmental plasticity are clearly adaptations to the physical environment. Jean Piaget was struck by the fact that water snails developing in pond habitats had elongated cells, whereas the same species developing in more turbulent lake conditions had more compact shells.[39]

Wing patterns on butterflies have been influential cases of developmental plasticity. Each wing is a monolayer of cells, the patterns being formed by local synthesis of pigments. In some species, little if any relationship has been found between genes and wing patterns, allowing enormous diversity of colour patterns.[40] This permits, for example, 'seasonal polyphenisms' where the shading of the wings, as they emerge from the pupae, shifts from tan in spring to red in autumn (as in the buckeye butterfly, *Precis coenia*).

Sometimes gross changes in morphology are involved. Some of the most striking examples are the castes in bees and ants in which developmental plasticity totally alters behaviour and physiology as well as anatomy. Locusts develop different physiologically and behaviourally distinct 'morphs' in response to current population densities. Likewise, the sex ratio in certain reptiles is known to be developmentally plastic, so that the embryo develops into a male or a female depending on local conditions, such as temperature, at the time. Metamorphosis in frogs and other amphibians involves the remodelling of almost every organ in the body, radical change in behaviour from filter feeder to pradator, and in locomotion and visual system. In some cases of developmental plasticity, the differences between the morphs have been so stark as to lead to them being classified as different species.[41]

Erica Crespi and Robert Denver point out that developmental plasticity is often mediated by the 'neuroendocrine stress axis' mentioned in Chapter 4. As we have seen, activation of it leads to internal physiological changes that improve immediate adjustment to situations.

But these can also have long-term consequences for development. For example, in humans, disturbances in the womb, such as nutrition, hypoxia, infection or chemical pollutants, can lead, through that axis, to accelerated organ (including brain) development. The paradoxical consequence of the disturbances, therefore, is an earlier birth. As Crespi and Denver put it, 'a hostile fetal environment in mammals accelerates development such that the individual can transition to a terrestrial environment where resources may be more plentiful.'[42]

In fact, many stressful effects in the parental environment are now known to be transgenerational, affecting children *not* experiencing the same stress, even across two or more generations.[43] This form of developmental plasticity, which arises from adjustments to the way that genes are recruited, has significant implications for studies of assumed 'genetic' effects on human development.[44]

Of course the most celebrated example of developmental plasticity is that in the brain. I will be discussing some of the consequences of this later: for the moment let's confine attention to the 'setting up' of the brain's basic cellular configuration. Canalised development seems to ensure that neurons in different layers in the cerebral cortex – the most recently evolved aspect of the brain – form different 'computational types'. But the 'wiring up' of these into specialised areas, with their specific response properties seems to depend upon context and experience. In one set of studies by Colin Blakemore and Richard van Sluyters, cats were reared in conditions in which they were exposed to lines of only one orientation (i.e., vertical lines or horizontal lines, or anything in between, so long as they were invariant for each cat). The cats developed a preponderance of neurons in the visual regions of the brain that were tuned to that *particular* orientation (as opposed to the full range normally developed).[45]

Such plasticity of development has been demonstrated spectacularly in other ways. Mriganka Sur and his colleagues surgically re-routed 'visual' connections from the eye in newborn ferrets away from their usual destination in the visual cortex, to what usually develops as the *auditory* cortex (i.e., processing auditory information). What has long been recognised as the 'auditory cortex' developed, in those animals, into a *visual* cortex instead (processing visual information).[46] Somewhat similarly, it has been shown that plugs of cortex transplanted from, say visual to somatosensory areas (responding to sensations of touch), develop connections characteristic of their new location, rather than of their origins. Finally, it has been shown how the functions of one area, surgically removed, can be taken over by another.

A host of new technologies, including high-resolution cellular imaging, genetic manipulations, fluorescent DNA probes and large-scale screens of gene expression, have revealed multiple molecular mechanisms that underly structural and functional plasticity in visual cortex. All this suggests, again, that, rather than being predetermined by a genetic 'code' for cortical functions, these area functions emerge from the dynamics of local spatiotemporal structure operating at critical times *in the course of* development.[47]

The induction of developmental plasticity appears superficially to be a response to singular cues. But, in reality, the structured responses of cells and tissues are themselves coordinated with the structural (spatial and temporal) information in the outside world as a way of predicting the future. This is, perhaps, seen most clearly in the developmental plasticity in many plants where factors can be more easily controlled and separated in experiments. For example, developmental decisions, such as flowering time, are based not on simple triggers, but on interactions among a number of different environmental variables. These include overwintering temperature, temperature during growth period, current length of day, and quality of light, which reflects the degree of crowding in the current site and so on.[48] In the case of wing shading in the buckeye butterfly, mentioned above, it was shown that both lower temperature and shorter daylight are factors in the developmental plasticity. But the development is sensitive to an interaction between them in that the shorter the daylight the greater the effect of the lowered temperature.

In some species of frog tadpoles, developmental plasticity, such as timing of metamorphosis, depends on the integration of information from multiple environmental sources.[49] In reality, an array of environmental factors acts 'through a complex interplay of direct sensation and indirect effects on behavior and feeding'.[50] Certain aphids have offspring that can develop as either winged or non-winged varieties. If parental aphids sense beetle predators in their locality, they produce up to three times as many winged offspring than they would normally. This allows dispersal to safer areas, but is not, of course a 'conscious' decision on the part of the mother, so much as a dynamic response to external environmental structure. For example, the proportion of each variety also varies with atmospheric factors such as carbon dioxide and ozone levels, and in ways that interact with the presence of parasites and predators. Under high carbon dioxide they produce more winged versions in the presence of predators. Under high ozone they produce more of the winged form in response to parasites.[51]

In each case, it's as if development is trying to predict the future environment from its current structure and, therefore, predict which phenotype would be best for facing up to it. As Robert Denver explains, far from a discrete trigger, 'Various combinations of biotic and abiotic factors can interact in complex ways to influence the growth and development of amphibians.'[52] Among the many factors interacting to influence metamorphosis, for example, he mentions temperature, concentrations of salts and gases, water acidity, change of light, tadpole density, depth of water, food density and predator presence, while none is independently decisive.

Studies of epigenetics in developmental plasticity have continued apace with some surprising twists and turns. One of these has been the accumulating evidence over the last twenty years of 'Lamarckian' type inheritance of acquired characteristics. This refers to the theory of Jean-Baptiste Lamarck who, in the nineteenth century, proposed an alternative theory of evolution to that of Darwin. He proposed that bodily features and functions could be altered during life, as with exercise and experience, and the modifications could be inherited by offspring. It is of course a conventional dogma, today, that the only modifications that can be inherited are those that occur accidentally in the genes. However, a number of mechanisms have now been identified through which the genetic material can be altered through environmental experience and passed on to the next generation.[53] Related to this is the large numbers of studies in humans and animals showing that the biological and behavioural effects of experience in a parental generation can be expressed in subsequent generations through tuning of epigenetic processes, as mentioned above. Even effects of experiences preceding pregnancy can be passed on to offspring.[54]

Obviously, such findings have deep implications for understanding human development and for methods aimed at unpicking the interactive processes involved in creating individual differences. Other extensions of the field are those that include developmental plasticity in physiological traits. For example, early exposure to certain bacterial toxins in infancy can alter hormonal and immune responses in later life.[55]

As might be expected, mechanisms for developmental plasticity are likely to be found in the signaling and gene regulatory networks described in Chapter 4. We have seen that signaling pathways can alter developmental pathways in a context-dependent manner. We also know of the non-linear dynamics in metabolic/signaling pathways, and in developmental processes, that can explain these effects.

We have seen that environmental effects can modify genes during life, as well as reprogram the epigenetics of how they are used. Moreover, modularity in developmental networks may provide a dynamic, self-organised 'toolbox' of developmental pathways that can be assembled combinatorially, to create novel phenotypes. Regulated changes in the recruitment of homeotic genes, for example, can cause one body part to transform into the structure of another body part. Although some of these effects may be deleterious to the organism, they suggest how rapid combinatorial modification of a general body plan is possible.

In developmental plasticity, then, we can see another solution to our mystery about the origins and evolution of complex structures. Species don't have to wait passively for a lucky mutation, and for its slow selection over countless generations, to be able to survive in changing environments. Their self-organising properties can actively create novel, and more complex, solutions far more quickly than that. Although always constrained to some extent by the genetic resources available we don't have to think of the 'evolvability' of species purely in terms of random genetic variation. Large changes in morphology and function can be achieved by alterations in the spatiotemporal timing of signalling and/or transcription networks, during development. Among other things, this explains why there is little relationship between species complexity and numbers of genes. For example, only around 1% of the human genome is different from that of chimpanzees, yet there are vast expression differences, especially in the brain.[56]

These are further examples of the way that dynamically structured responses in organisms can be active contributors to evolutionary futures and not just be passive subjects of blind mutation and selection. But it's also important to recognise the dynamic reciprocity with the environment. Novel morphological or behavioural forms can arise through developmental plasticity, and these can permit habitation of more complex environments. They can, in turn, lead to rapid species diversification.[57] However, those new forms also change the environment. For example, they are likely to bring about alterations to species interactions and food chains, that can, over time, restructure the living world and also alter trajectories of evolution. Recognition of such integration of developmental plasticity and ecological context has helped form the new sub-discipline of 'ecological development', or eco-devo.[58]

Developmental plasticity is, then, another illustration of emergence of higher-level structures and regulations from interactions among more basic components. It turns development from a purely passive assembly of parts, with a specific end, into an important system of adaptability in its own right. This is why it is not an isolated phenomenon. The huge significance of developmental plasticity for evolution, especially in the human lineage, has recently been emphasised by many authors.

Epigenetics 3: Lifelong plasticity

All the examples of development mentioned so far reflect responses to experiences occurring before the end of physical maturity, and, therefore, 'fixed' before adulthood. Development, as described so far, does a positive, constructive, job in improving the adaptations of species, reflecting another leap in the evolution of complex systems. But even that is limited. Once a snail has developed a pointed as opposed to a flattened shell, it cannot change back. Once a barnacle has developed a bent as opposed to a straight form, this cannot be reversed or altered or extended. Once some tissue inflammatory tendencies have been established in early life, they don't tend to change back.

On the other hand, environments continue to change on macro- and micro-scales throughout life. The 'one-shot' developmental plasticities discussed so far cannot cope with that. What is required for survival in such conditions are living structures and functions that can continue to change *throughout the life of the individual*.

James Baldwin was the first to suggest plasticity as a 'method' for dealing with environmental novelty and adaptability throughout life.[59] There is abundant evidence for the existence of such continuously adaptable characters. Simple examples include tanning in the sun; growing protective skin calluses at points of friction; increasing muscle bulk and strength with exercise; and changing hair colour and thickening with the seasons. General physiological fluctuations in response to rapidly changing environments have also been described as lifelong plasticity.

Lifelong plasticity seems to have emerged quite early in evolution. Certain plankton species really need to be transparent for protection against fish predators, yet also need protection against UV light. Accordingly, they can reversibly develop body pigmentation as and when needed.[60] Whereas the skin patterns of most reptiles is fixed at the end of development, that of the chameleon retains a lifelong changeability. Bird song control systems exhibit seasonal plasticity in many species. These include dramatic volume changes of entire brain regions in response to photoperiod (length of day) and its impact on circulating levels of sex steroids.[61]

Sometimes, this plasticity can involve spectacular body changes, as can be seen routinely in certain coral reef fish. 'These fish live in spatially well-defined social groups in which there are many females and few males. When a male dies or is otherwise removed from the group, one of the females initiates a sex reversal over a period of about two days, in which she develops the coloration, behavior, and gonadal physiology and anatomy of a fully functioning male.'[62] A wide variety of such 'inducible' phenotypes, responding to environmental signals across the lifespan, are known. In some social insects, 'soldier' castes may be induced by collateral changes in their prey, such as the appearance of more defensive phenotypes in aphids. These changes are akin to rabbits changing to porcupines because wolves have been smelled in the woods.[63]

An example of lifelong plasticity we tend to take for granted is that of the immune system. It works by binding the foreign protein (antigen), secreted by a microbe, with a structurally matching antibody, produced by a type of white blood cell (B cells). The bound antibody then triggers rapid division and differentiation of the B cells to produce massive quantities of that particular antibody. They do this by utilising the signaling and gene regulatory networks described in Chapter 4. In this way entirely novel antibodies may be produced throughout the life of an individual. The individual 'profile' of antibodies continually develops according to life history.

Evolution of behaviour and cognitive systems

More generally, though, in tracking lifelong, *rapid* environmental changes, behaviour – and, more important, the brain and cognitive regulations governing it – becomes paramount. Behaviour, in fact, probably emerged as an instrument of lifelong adaptability very early in the course of animal evolution. It is changes in behaviour that have, above all else, determined the course of evolution of animals over the last hundreds of millions of years.[64] Yet poor understanding of the evolutionary status of behaviour – and especially of how it is regulated – creates a major hiatus in our understanding of those systems.

One cause of this hiatus may be that behavioural biologists and psychologists have tended to focus on relatively 'closed' or 'instinctive' behaviours. These include the range of mating, nesting, territorial, migration and hibernation routines, familiar in popular literature and the media. This focus is understandable in that such behaviours present a uniformity and repetitiveness convenient for scientific observation and analysis. But they represent developmentally canalised sets of narrow, well-adapted responses to well-defined, recurring stimuli. They are not typical of evolved behaviours and their regulation in rapidly changing environments.[65] Although study of them over the last fifty years, or so, has contributed to an exciting biology of behaviour, it neglects a crucial stratum of higher behavioural regulations as evolved in cognitive systems.

Although physiological and epigenetic regulations do a pretty good job of dealing with environmental change at a certain level, the rapidity and variability of change we are now talking about is in a different league. 'Constant novelty' becomes the order of the day. As described before, such change can only be rendered predictable if it is based on some deeper informational structure. But *making* it predictable requires a whole new layer of regulations that can abstract that structure on a continually renewable basis – *and*, of course, construct motor actions *de novo* to ensure the most appropriate response to the present situation.

As animals have been forced to inhabit more changeable, less immediately predictable niches, the need to capture this deeply structured information has increased enormously. This was the basis of the evolution of complex nervous systems. They have evolved to abstract such informational structure through rapid modifications of nerve cell connections in large-scale networks. Eventually, the epigenetic developmental systems described so far became the basis of lifelong plasticity underpinning cognitive systems. This is why lifelong plasticity is the operational *modus vivendi* in the more recently evolved parts of the nervous system. It explains the mushrooming size of cerebral cortex in the evolution of vertebrates. 'The interaction of individual animals and their world continues to shape the nervous system throughout life in ways that could never have been programmed. Modification of the nervous system by experience is thus the last and most subtle developmental strategy.'[66]

This strategy, and its coordination with complex forms of environmental structure, is broadly the subject of the next five chapters. In the meantime, it's important to acknowledge how development, originally a means to an end, has, in cognitive systems, become the paramount strategy for lifelong survival. Let me prepare the ground for that later discussion with a quick look at some principles of brain development.

Brain network development

As mentioned above, brain cells and their elaborate synaptic networks appear 'almost magically' during a brief period of embryonic development. Starting as a fold of cells in the early embryo, it soon gives rise to a vast diversity of other cells that migrate widely. The whole process is self-organised through extensive signaling and gene regulatory networks.[67] As in other organs, the process exhibits both specificity (i.e., canalisation) and plasticity at different times during development.

The complexity of the process is truly remarkable. Brain cells (neurons) are produced in the embryo/foetus at a rate of up to 250,000 a minute. These progenitor cells have to differentiate into the many types of immature neurons in the embryo/foetus and find their correct positions, often in precise layers, at the correct time. In many cases highly specific positioning of cell bodies is only found by migration over long distances in the three-dimensional structure of the brain, very often in specific strata, as in, for example, the six-layered structure of the cerebral cortex. Then they have to produce the processes (axons and dendrites – see Figure 5.1) through which they interconnect.

This wiring up of the nervous system is no easy task among what, in humans (the extreme case), consists of around 100 billion cells and trillions of connections. Axons, carrying signals from a cell of around a thousandth of a millimetre in diameter, may travel up to a metre to reach other cells in precise locations, often on distinct layers. There, the terminal branches must end up in the right order at the right time on the right parts of the web-tree of receptor sites on the dendrites, on the target cell. They may be doing this alongside dozens or hundreds of other axon terminals from other neurons.

Neuronal progenitor cells get 'wired up' correctly by communicating intensively with each other through chemical signals, guiding the axons along the correct pathways.[68] Target neurons seem to produce surface receptor proteins as 'labels' for approaching axons, while other 'labels' are spread out as molecular gradients, attracting some axons and repelling others, often over long ranges. Large families of these guidance systems have now been identified.[69] The branching tip of the axon growth area – or growth cone – possesses remarkable abilities to detect, not only a rich variety of molecular directional cues within the molecular soup it traverses, but also their spatiotemporal pattern. The timing at which the axon cones become 'competent' seems to be crucial in determining their responsiveness to the guidance molecules, and thus their destinations.[70]

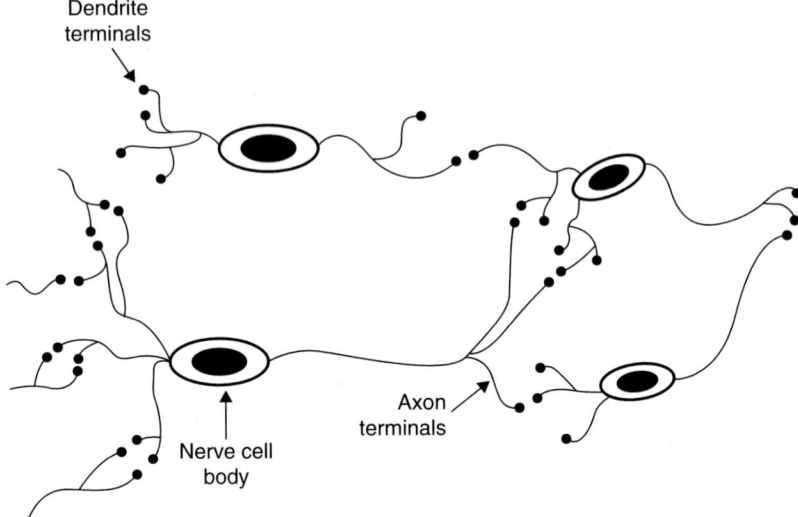

Figure 5.1 Getting connected. Axon terminals, guided by molecular signals, find target cells. There, development of dendrite terminals is promoted by structured spike firing along the axons

In mammals, especially primates, much of this occurs in the uterus before precise sensory experience. But it is also known that spontaneous signaling activity between neurons is needed for connections to begin to form. Injection of drugs that inhibit such firing much reduces the formation of connections. More importantly, correlated inputs from sensory centers (such as the retina) to internal centres, and feedback from higher cortical centers, seem to play their part. Although it has been known for decades that light experience is important for development of functional connectivity in the visual system, it seems to be *patterned* (i.e., spatiotemporally structured) light that is crucial. Confining visual experience to 'white noise', presenting all light frequencies without the patterned quality, retards development of connectivity.[71] This prerequisite also seems to apply in the auditory system: exposure to white (uniform) sound during early development seems to severely disrupt its proper development.[72] Far from the brain being a 'blank slate', therefore, it has already been geared up for the abstraction of informational structure long before birth.

This initial wiring up can be plastic, as mentioned earlier, and the connectivity can be revised, depending on changes in activity both locally and in other areas. But it is the *lifelong* plasticity that defines the brain and its cognitive functions. It is a system in which developmental plasticity, previously confined to early life, has been extended so that development itself has become the prime means of adaptability in adulthood. However, it is also important to remember how that very adaptability, and its effects on behaviour, also changes the environment in which the system lives in ever-increasing ways. This creates still further challenges for developmental systems – and so on, in a spiral of complexity. Development, for example, serves as a crucial springboard to the evolution of further complexity in systems. This is why we really must take development seriously.[73]

Meanwhile, it is worth emphasising that complexity of developmental systems, and their integration with environments, isn't captured in common descriptions such as 'a mixture of genes and environment', or 'nature and nurture'. Instead, authors increasingly speak of a constructive dialogue between internal systems and the dynamics of the environment. In an important sense, those dynamics become part of the fabric of living things. This will be illustrated more vividly when we turn to aspects of cognitive functions.

6
Intelligent Eye and Brain

It is a feature of evolution that the emergence of one new trait almost invariably affects, or makes demands upon, several others. So the evolution of bodies that move increasingly rapidly and fluently increased selection pressures for other forms of support. One of these is a range of sensory organs, for registering the state of the outside environment, and a brain, for interpreting this information and coordinating responses to it. Such support became of paramount importance as animals were forced to inhabit more changeable parts of the environment – a world teeming with objects, some as obstacles to activity, some to eat, some to be eaten by, some to hurt, and some to protect or hide behind.

This was a world in a new league of changeability from the flowing concentration gradients experienced by the first primitive organisms. Its growing importance is almost directly reflected in the evolution of the nervous system: from a thin 'nerve net' and singular sense organs in early multicellular species such as the Hydra; to round worms that have concentrations of around 300 neurons to form a primitive brain; to insects with three times as many neurons; to mammals, primates and humans with up to ten billion neurons and over 50 trillion connections between them.

The questions to be tackled in this chapter will be: what is this burgeoning system actually for? How does it work? And what does it tell us about the evolution of complex systems in general, and of the mind in particular? After all, the brain is thought to be the seat of intelligence in most species. But how has it made intelligence more complex? We still often resort to rather simple metaphors of the brain such as a calculating machine or a computer, or a factory with its divisions of labour and hierarchy of departments, all under the control of a 'central executive'. How valid are they?

All this creates much to discuss, but for parsimony, in this chapter, I will focus on vision and a fairly general vertebrate brain (although most of it can be extended to other sense modes). In all cases, we must assume that the whole system is the answer to an evolutionary problem. But it's important to ask, what, exactly, is the problem? And what kind of answers do we really get?

Seeing isn't easy

In spite of some remarkable technological and empirical advances over the last fifty years, the answers have been elusive. In a recent article entitled *The Unsolved Mystery of Vision*, Richard Masland and Paul Martin say that, 'Vision looms large in neuroscience – it is the subject of a gigantic literature and four Nobel prizes – but there is a growing realization that there are problems with the textbook explanation of how mammalian vision works'.[1] They go on to present a portrait of a field that, they say, is 'stuck'. So, once again, we discover that the mystery surrounding the evolution of mind is actually a series of associated mysteries. Masland and Martin do, however, point to the possibility of new foundations, and hope for the future. It's those foundations, and what we might learn from, and build upon, them, that I want to end up discussing. But first let us look more closely at the problems.

A simple, and still popular, view of vision is that the light-sensitive part of the eye – the retina – just 'picks up' the main features of objects in topographical form and passes them on to the brain, which then puts them back together in their original configuration. In this way visions of the physical world seem to follow each other in an orderly, stable succession: a three-dimensional, seamless cinematography in the head. Alva Noë calls this the 'snapshot' view of visual experience.[2] For most of us, most of the time, the process works so brilliantly that we hardly ever stop to ask how it's done. Let's take a look at how it has been studied and what has been revealed.

Feature detection?

In that popular view, objects are experienced by virtue of the reflected light received by the eye as an impression or image – a 'snapshot', like a photograph. The sensory cells, in large numbers in the retina at the back of the eye, each pick up a tiny part of that image as tiny light spots. Light energy falling on photosensitive molecules causes each cell to discharge an electrochemical signal down its axon. The discharges converge on the ganglion cells (see below) where they are received as primary 'features' such as lines or edges. Axons from the ganglion cells then converge in the optic nerve, so that the whole snapshot is passed on down the optic nerve, as a battery of discharges, to the brain (this is sensation or sensory reception). The brain re-associates these features through a series of further convergences into an image of the original object (this is perception). This image can then be handed over for other functions to be executed, such as recognising, classifying, memorising, thinking about it, action planning and constructing a motor program (this is cognition).

The view has been much encouraged by studies of response properties of neurons themselves, in the various way-stations in the visual system. This has been achieved by inserting a very fine microelectrode through the eye or brain into or near a neuron in anaesthetised animals, then flashing a range of visual features, such as light spots, lines or bars, into the eye. When the neuron responds, its discharge can be registered through the electrode, and we can take it that the neuron is sensitive to that particular feature. Such studies have seemed to confirm that neurons in the visual tract are specialised to detect one or other of those features.

This 'feature detector' theory, deriving from experimental work of Steven Kuffler, Horace Barlow and others in the 1950s, thus reinforced the idea that the response properties of neurons exactly correspond with specific, stable, sensory impressions.[3] Implicit, of course, is that the task of the eye-brain system is the internalisation of spatially stable, temporally static, object form – the counterpart in the nervous system of the Darwinian view of a stable, static environment in experience in general. Let us briefly consider why the view has encountered problems: why Masland and Martin go on to claim that 'the standard view of visual system function is in trouble from stem to stern.'[4]

Sensory reality

There are many reasons why feature-detection alone cannot create the image required. One of these is that the world is populated with objects that are three-dimensional: a three-dimensional object is collapsed as a 2D image on the retina, and the same 2D image can be created by a number of different objects/features. The 2D image is also grossly distorted because the back of the eye is hemispheric (imagine watching a movie projected onto a bowl-shaped screen). Moreover, the receptor surface is not a continuous 'sheet' (like a camera film) but an aggregate in the form of millions of light-sensitive cells. So the visual field is detected in the form of a moving cloud of light spots. For all these reasons, percepts cannot be exactly determined by the sensory input, but require some kind of inferential construction as well. Irving Rock called it perceptual 'problem-solving', or 'intelligent perception'. Like the other intelligent systems we have discussed so far, this means 'going beyond the information given'.

Another problem is that the image of any single object is never still. It is under constant spatiotemporal transformation as we move around it, or it moves to different places at different distances (with changes in apparent size), rotates, passes behind other objects and so on. In addition, the brain must distinguish those motions from the motion of the image on the retina due to eye movements and those due to personal bodily movements. And there's another problem. It takes 30–100 ms for sense receptors like those in the retina at the back of the eye to respond and pass the signal on. This is slower than most motion, so by the time we have perceived an object in motion it is no longer exactly where we now 'see' it. This makes actions such as catching a ball more difficult than you think, not to mention catching prey or avoiding predators, like our ancestors did: actions have to be 'ahead' of vision, as it were.[5] Finally, a single scene may contain dozens of objects so that what we constantly face is *'an onslaught of spatio-temporal change.'*[6]

In considering the nature of visual experience, William James was closer to the reality in 1892 when he famously referred to the 'blooming, buzzing confusion' of the sensory input experienced by the infant.[7] It all supports Donald Hoffman's view that, in making any sense at all of such experience, we are all, in fact, visual geniuses naively unaware of our rich talents.[8]

The structure of the visual environment

Here, again, we seem to encounter the problem described in previous chapters: preconceiving how a system works and then assuming the (impoverished) environment that satisfies it. One irony of the difficulty, here, is that it has long been known that vision originally evolved to detect movement, and to provide sensory guidance for animal motion.[9] In fact, we know that motion sensitivity in the retina is extremely acute. Even tiny motions, less than the diameter of a retinal receptor, can be detected.[10] This sensitivity is ubiquitous in animals with eyes, as seemingly basic as detection of light and dark, and 'may be the oldest and most basic of visual capabilities.'[11] Perhaps, then, the evolution of vision reflects a greater complexity of experience than we have so far imagined. The fact is, that motion, by definition is change. And, as described in Chapter 3, a changing environment presents far more complex structure-for-predictability than a static world. This is particularly so in a world of objects.

As argued before, it is the structure beneath the complexity that yields predictability. What a sensory system particularly needs, then, is the structure that comes from motion. This has been confirmed directly and indirectly many times. For example, deficits in motion detection are quite devastating. This was strikingly demonstrated in one patient who had suffered brain damage from a stroke. The patient had great difficulty pouring coffee into a cup. She had some perception of the cup's shape, colour and position on the table, and was able to pour the coffee from the pot. But the column of fluid flowing from the spout appeared frozen, like a waterfall turned to ice. Without perception of its motion the patient allowed coffee to over-pour into the cup and spill over the sides. More dangerous problems arose when she went outdoors. She could not cross a street, for instance, because the motion of cars was invisible to her: a car was up the street and then upon her, without ever seeming to occupy the intervening space. Similarly with people: they were 'suddenly here or there but I did not see them moving.'[12]

It has also been demonstrated many times that the 3D shape of objects in motion can be recognised better and faster than static objects. In a classic experiment by Hans Wallach, over fifty years ago, participants were shown 2D shadows of a wire-frame figure on a screen, and asked to report their perceptions. They reported that the image appears flat when the wire frame is stationary, but 'pops out' in 3D depth as soon as it is rotated.[13]

Another popular demonstration indicates how whole figures can be recognised from a few light points moving with appropriate coordination, but not in static presentation. One of the best illustrations is the point light walker (Figure 6.1).[14] There are many other examples. The structure in motion has been crucial for the evolution of mechanisms for perceiving change. Indeed, most animals fail to respond to a perfectly motionless object, as if they cannot 'see' it. If a dead fly on a string is dangled motionlessly in front of a starving frog, the frog literally cannot sense this meal to save its life. In humans it has been shown that perception of a retinal image held perfectly still on the eye (by using a kind of contact lens) quickly fades and 'disappears'.[15]

Intelligent Eye and Brain 95

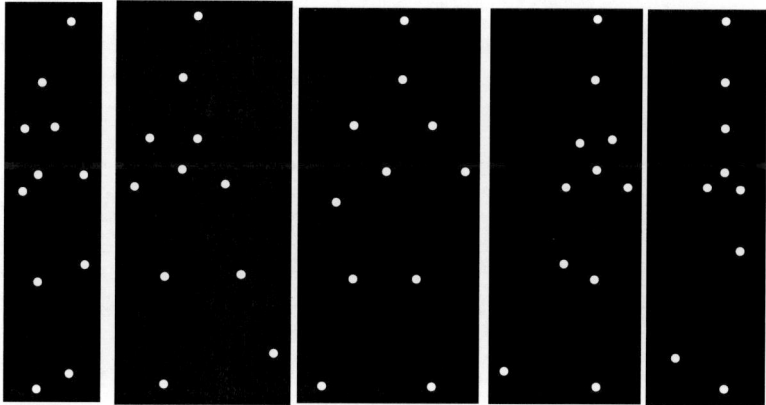

Figure 6.1 A sequence of stills from a 'point light walker'. Volunteers find the images unrecognizable when presented individually, on a computer screen. But presentation as a sequence at normal speed evokes almost immediate recognition of a person walking

What kind of structure?

Motion reveals much about some superficial attributes of objects, such as relative distances from the eye, different surfaces and so on. But there's much more than that. When we move around a book on a table, say, something invariant about its 'bookness' persists in spite of radical changes in appearance. As Kevin O'Regan and Alva Noë say, there are systematic changes in the received visual patterns that immediately distinguish one object from another.[16] But what exactly is the nature of those patterns?

A widely accepted idea is that the structure exists in the form of statistical associations between the variables-in-motion. As Horace Barlow put it, 'The visual messages aroused by the world around us are full of very complicated associative structure.' Indeed, he has demonstrated that single-cell recordings in retina and brain are sensitive to associations in repeated stimuli, and suggests that the brain may use 'higher order' associations.[17] Joseph Lappin and Warren Craft suggest that structure has to mean spatial relations between points emanating from a light source. They show that it is the change of these relations over time – in the course of motion – that reveals the structure: 'the image changes produced by motion are visually coherent – correlated – yielding spontaneous organization of the spatially distributed changes within and between features.'[18]

In other words, the spatial structure of objects is defined by some sort of covariations in the *spatiotemporal* changes intrinsic to them. The next question is: How has the eye-brain system evolved to handle them?

Assembly of features?

The retina consists of layers of cells about 0.5mm thick lining the back of the eye. The light-sensitive cells are the rods and cones, so called from their shapes (Figure 6.2). Numbers and proportions vary in different species, but the human retina has up to 120 million rods (sensitive to white light) and about 7 million cones (maximally sensitive to one or other of red/green/blue light). In response to a stimulus, light-sensitive protein pigments in the rods and cones trigger a sequence of molecular responses, the phototransduction cascade. This is passed on subsequently as an electrochemical signal to other cells in the layers through their connections or synapses (Figure 6.2).

After the signal has passed from the receptors to the bipolar cells, it is propagated by a series of 'spikes' of electrochemical activity (the nerve 'impulse' or 'action potential') along the axon to the ganglion cells. Impulses from the ganglion cells are then passed along the axons of the optic nerve to the lateral geniculate nucleus (LGN) and other parts of the mid-brain. From there they go to the various levels of the visual cortex and other parts of the brain. All that occurs at phenomenal speed. There is useful information in the cortex within one-sixth of a second, while actual motor response (e.g., speech or pointing) starts only a couple of hundred milliseconds later.[19]

As mentioned above, fine glass electrodes inserted into or near single retinal ganglion cells (RGCs) in anaesthetised animals have been used to glean information about their responses to light stimuli. A variety of simple patterns are projected onto a screen in front of the eye. Whatever pattern a given cell responds to was deemed to be the information it 'codes', or its 'receptive field' (RF). For a while, the responses seemed to be clear and unambiguous. The RGCs seemed to be responsive to light spots, or edges or bars moving across the visual field, some only in a particular direction (these RFs vary with species, as we shall see later). Similar recordings from the LGN and parts of the visual cortex, seemed to confirm the assembly line mentioned earlier: light 'spots' become associated as edges, these as more complex object parts and so on, in the primary visual cortex, while 'hypercomplex' cells in the 'higher' cortex seemed to respond preferentially to complex, compound shapes, such as whole objects or faces. This has led to the idea that

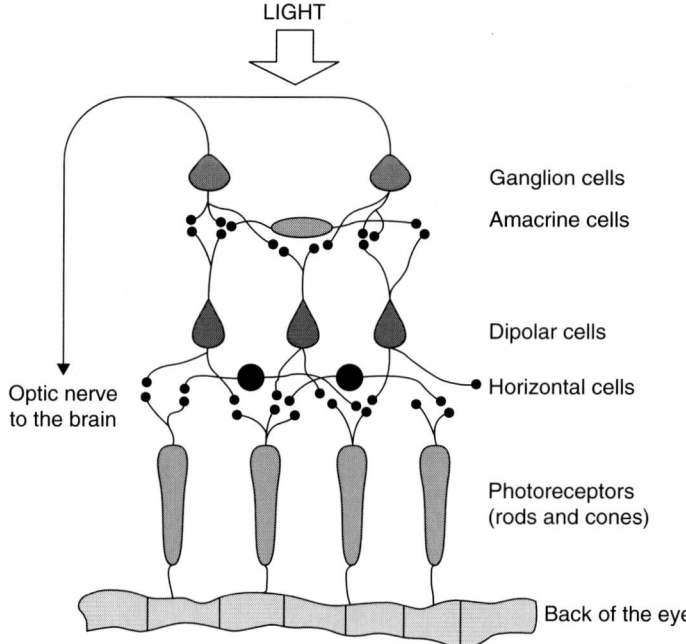

Figure 6.2 Schematic diagram of retina and cell connections

such cells 'represent complex shape information' for visual recognition of objects.[20] The idea, of course, reflects the view that the purpose of vision is to reconstruct an exact 're-presentation' – the 'what' and the 'where' – of the outside world.

Revision of the feature detector doctrine

The problem with the traditional feature-detection approaches is that they tell us only how a given cell responds to the static, minimalistic stimulus that the experimenter happens to choose to 'flash' at it. In addition, the animals are anaesthetised and restrained in far from normal circumstances. This does not tell us much about responses of cells in more natural contexts. Accordingly, the full contributions of these neurons to visual scene perception remain vague. As Thomas Albright and Gene Stoner say, the standard interpretations reflect an 'elementism' based on an old, non-dynamic view of visual experience.[21]

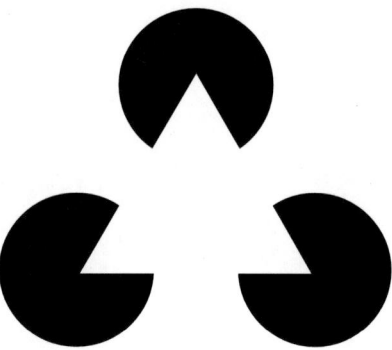

Figure 6.3 Kanizsa triangle

Besides, we now know that the feature-extraction model is inconsistent with other observations suggesting that cells are not just firing to features present. For example, the point on the retina through which the optic nerve passes creates a 'hole', or blind spot, in the visual field, devoid of receptors. Yet the fact that we are not conscious of that hole, and don't seem to see it, suggests some compensatory mechanism based on other kinds of information. Indeed, there are neurons in the visual cortex that actually respond to light stimulation on the blind spot. Cells are also known that show responses to aspects of stimuli not actually visible, as in occluded aspects of overlapping surfaces and objects, a phenomenon sometimes exhibited as visual illusions. For example, neurophysiological studies have identified cells in the visual cortex that respond to the parts of the Kanizsa triangle (Figure 6.3) where there are no lines, although we have the illusion that there are.[22]

All this suggests a constructive process based on more than a linear convergence of primitive features. These revisions have been confirmed by studies of RFs using naturalistic stimuli, as well as new techniques for recording from multiple neurons simultaneously.

Interactions in the retina

The upshot of these studies is that image construction from ambiguous data is indeed based on deeper, structural information, just as life in the real world demands. As Albright and Stoner say, 'Contextual influences on perception are necessarily manifested as interactions, rather than passive associations, between sensory, behavioral, or cognitive elements.'[23] But this implies that the cells of the visual system need to be interested in something other than independent light spots or features.

Indeed, the very rich interconnectivity within the layers of the retina suggest numerous ways in which the myriad input channels effectively 'talk' to each other, and mix and exchange information (only sparsely represented in Figure 6.2). What they seem to be talking about is the correlational structure of the light variables, over thousands of receptors, at probably many levels (i.e., higher-order correlations, or mutual information). It is that stimulus structure that is eventually heeded in the retina.[24] Once assimilated from experience it can be used to generate predictions about the identity of future – inevitably partial, fuzzy and novel – visual inputs.

This has been shown in many ways, using more advanced methods. Rather than using the classic static stimuli, Frank Werblin and colleagues recorded from large numbers of RGCs, *and* from cells in the other layers of the retina, simultaneously. A small bar of light was then shone across a row of receptors in succession (in either retinal slices or perfused whole retina in a dish). By playing back recorded signals from the cells, the researchers were able to reconstruct what they call 'spatial-time' maps of responses across the row of cells at successive moments, in the different layers. These responses showed that there are excitatory and inhibitory interactions across and between layers such as to define about a dozen types of space-time 'representations' at the RGCs. Rather than static features, each property reflects a specific type of change or transformation in the visual scene, being integrated in ganglion cells, and output to the brain.[25]

In other studies, effects of movement context, reflecting the registration of higher-order structure, have been observed. Bence Olveczky and colleagues discovered RGCs in both rabbit and salamander that respond to motion, but only if the background moves in a different direction (otherwise they are suppressed).[26] In another research, it was observed that the directionally selective ganglion cells also respond to the temporal patterns of illumination occurring nearby. Therefore, 'some properties of ganglion cell responses will be revealed only by an analysis of higher-order sequences in the stimulus set. If higher-order sequences are rare stimulus events, why do we need to care about them? The reason is obvious: They are not rare events in the natural world – they are the very common sequences generated, for example, when any visual stimulus moves continuously across the retina.'[27] All this already suggests that the visual experience handled by even the early visual system is more complex than we previously envisaged: but there is more.

Assimilating informational structure

The reason that motion sensitivity is so important should now be obvious. Changes over time provide the crucial informational structure for enhancing predictability about both the identity of inputs and their future positions. Statistical analysis even of pair-wise pixels of light moving on the retina show that there is structure present. As already mentioned, any light 'fragments' reflecting from a naturally moving object do not change independently; their spatial and temporal dispositions are highly correlated.[28] Even simple features experienced in natural conditions will be full of spatiotemporal covariations (correlations) between constituent light points. In fact, when the moment-to-moment changes in multiples of sample light points like this are measured there is evidence of considerable deeper structure.[29] This is the kind of higher-order information that distinguishes a straight line from other structures such as curved lines. Given the predictability afforded by such structure it is hardly surprising that this is really what the system is 'looking for'.

More striking support for such ideas comes from two other sources. First, from research on response properties of neurons to naturalistic stimuli, rather than isolated and simplified features. Second, by recording from multiples of cells simultaneously, rather than independent cells. These attempt to relate cell responses directly to the structure in natural experience.[30]

In one study Jarpo Hurri and Aapo Hyvarinen analysed videos of natural scenes by breaking them down into frames (or small patches of the total frame) with tiny time delays of 40–960 milliseconds. They produced evidence that responses in cells, even to simple features like lines, is dependent upon 'higher-order correlations' among inputs, requiring time monitoring – that is, non-linear temporal correlation. As they say in a related paper, a more 'generative model' is needed in which 'hidden variables' can be used to make inferences about the underlying real world.[31]

In another study, salamander retinae were exposed to movies of natural scenes while responses were recorded from tens of RGCs simultaneously. It was found that the joint signals send a lot more information than the sums of independent signals would imply: in fact, the equivalent of a tenfold 'over-representation' of visual information.[32] There is much other evidence that important structural information about visual input is missed in single-cell analysis and that neuron's responses critically depend on the higher-order statistics of its inputs.[33]

Dynamics of visual information

We can now be fairly sure that the message provided by sensory systems to the brain must arise from the self-organising activities of the network as a whole. The sums of independent spikes from assumed independent features will not capture the deeper structure needed to make sense of the world. As mentioned above, the signals inputting the RGCs are themselves clearly products of complex interactions among horizontal and amacrine cells. The fact that the retina contains over 100 million photoreceptors but only one million ganglion cells indicates that it must perform significant data compression within the network. We don't know exactly how to describe the message sent from the ganglion cells, except that it must be some aspect of the deeper structure.

These dynamic processes of nerve networks can be simulated in computers. It has been shown how model networks involving plastic synapses can easily 'learn' patterns of correlations in inputs and capture their spatiotemporal structure: 'the spatiotemporal structure of the afferent input is imprinted into the learned synaptic efficacies.'[34] As mentioned in Chapter 2, such structure can only be assimilated by recording the statistical interaction parameters describing it. In dynamical terms, this will involve breaking previous equilibria with the system of interactions evolving to new basins of attraction in which the synapses attuned to those parameters grow stronger than those responding to the uncorrelated background. Inputs in which similar structural parameters are detected will be drawn towards such attractors, which then respond with a 'fuller' stimulus output.

In other words, we are learning that perceptual recording of the world teeming with objects in perpetual motion is not of singular, independent data points. The message is of a much more abstract, or deeper, one than sets of discrete feature images. What needs to be captured in this structural economy of vision are the deep spatiotemporal parameters emanating from the external objects and background environment. As we shall see below, the parameters probably form attractor basins at various interacting levels. The overall attractor landscape is what defines the invariant qualities of 'bookness' – and so on with all other familiar objects.

By attuning connectivity to interaction parameters, rather than independent features, the networks can more easily interpret novel data, yet permit an enormous variety of expressions for forward transmission. Compression of data into statistical parameters, rather than independent features, probably explains why there can be many fewer ganglion cells (and optic nerve fibers) compared with receptors. There is no loss of information ($A=B^2$ is a lot more economical than a long list of pairs of specific values). However, it is also probably the case that sensitivity, and response, to such parameters can only be achieved by dynamic states within the retinal network close to criticality and expressing occasional chaotic dynamics.[35] Non-linear dynamic systems are more interpretative and creative in extracting predictability from uncertain inputs and making fruitful responses to them. Evidence for the existence of such dynamics in the retina has been slowly accumulating.[36]

Note that, although I am concentrating on vision, similar principles seem to apply to other sense modes. For example, it is known that receptors in the cochlea of the ear, transform sound (in reality, already a spatiotemporal acoustic stream) into spatiotemporal response patterns sent along the auditory nerve.[37] Tactile senses, too, are able to abstract the deeper statistical structure from sequences of vibration.[38] With smell, it is more obvious. As Hong Lei and colleagues explain, 'An animal navigating to an unseen odour source must accurately resolve the spatiotemporal distribution of that stimulus in order to express appropriate upwind flight behavior.' They found that hawkmoths were sensitive through their antennae to the patterns of odour 'plumes' caused by air turbulence, and used them to get to the odour source.[39]

What the brain is for

In spite of enormous technical and research advances over the years, deciding what the brain is for has been surprisingly difficult. But again, the question we need to keep in mind is: what kind of problem is the brain an evolutionary solution to? This seems a good way to enquire about its functions. A subsidiary question will be why has the brain increased in size so much in evolution? As above I will be concentrating on vision. And again, I will be challenging the linear, feed-forward sequence of processing, such as feature-detection and compilation, as in a perceptual assembly line.

Most of the neurons in the brain share the basic features of a cell body with extensively branched dendritic trees at one end and a long single axon at the other. The dendrites receive signals from other neurons and the cell body responds with impulses along the axon that may or may not be extensively branched. Each neuron may receive inputs from thousands of others, and also send signals *to* thousands of others through its axon branches, some nearby, others at possibly great distance. Many neural models attempting to account for 'higher' functions such as object perception, conception and cognition have been proposed. The idea that the brain serves as a kind of calculator or computer has been around since the eighteenth century. Until recently, however, almost all assumed a hierarchy based primarily on the convergent, feeding-forward, of independent images.

We have just seen how the function of networks in the retina seems to be at least to start the abstraction of spatiotemporal structure in experience, rather than detection of stable, and probably mythical, features in the usual sense. The structure in natural experience is what affords predictability, and it is spatiotemporal in form. Computational brain scientists have been arguing recently that it is distinguishing just such 'structure in motion' that the brain is particularly good at. I will pull out the implications of such a view for understanding cognitive functions in a later chapter. But first let us look at what happens to the signals from the retina.

Signals from the retina pass, via the optic nerve, to the lateral geniculate nucleus (LGN), the superior colliculus (SC), and a couple of other centres, all in the evolutionary 'older' part of the brain called the thalamus. Although reciprocally connected, these two centres have distinct functions. The superior colliculus is involved in the control of eye movements, and in correcting for disparities between impressions of movement in the outside world and those created by self-movements. All sensory signals are also transmitted to parts of the mid-brain involved in emotion, feelings and monitoring of inner body states. The LGN, however, is the dominant target for retinal signals in more evolved mammals, particularly primates. It was naturally thought that understanding what the LGN does would be an essential first step in characterising the processing of visual information.

For a long time, in fact, the LGN was considered to be just a way-station in the transmission of visual signals into the 'real' brain. Its function just seemed to be that of passing on detected features from the retina to the visual cortex. This was suggested by both its position in the visual tract, and the fact that its neurons are arranged 'retinotopically', reflecting an orderly mapping of axons from RGCs in the retina. Initial single-neuron recordings also indicated RF's very like those in the retina. All this seemed to reinforce a feature relay model.

That the LGN is more than a mere way-station for simple feature traffic, however, is suggested by the fact that, of the several million axons terminating in the LGN, only about 10% originate in the eye. The rest come from several areas of cerebral cortex and a variety of subcortical regions.[40] It seems reasonable to ask, if the LGN is just passing on features picked up in the retina, what are all these other inputs for? Indeed, it has become well known that, much like cells in the retina, those in the LGN, are not islands with singular inputs and outputs. Rather they respond as part of a network to the wider context of change.[41]

Original RF studies had focused exclusively on the spatial aspects of response properties: that is, the shapes and locations of light sources, as with the feature-detection theory. But then the sensitivity of LGN neurons to temporal aspects of stimuli was shown.[42] Just as in the retina, more sophisticated recording techniques have permitted mapping of the spatial and temporal response patterns of LGN cells jointly. These show that the spatial structure of response of any particular cell itself changes over time (may become stronger, weaker, or change shape). Rather than reflecting a fixed feature, what they respond to can only really be described as space-time structures, like a movie. 'In fact', Gregory DeAngelis and colleagues say, 'when considered in the space-time domain, most cells in the geniculocortical pathway exhibit striking dynamics.'[43] The idea of fixed receptive fields is not really an accurate one.

This general picture has been confirmed, with the use of natural scenes as stimuli. For example, Garrett Stanley and colleagues showed that LGN neurons respond much better to natural scene stimuli than to traditional artificial stimuli (spots or bars of light projected onto a screen). This could be because the former have spatiotemporal correlational and non-linear structure.[44] They also showed that the typical LGN neuron response consists of two phases, a kind of 'wake-up' burst, signalling the presence of an object in the visual field (to alert higher centres, modulate attention and eye movements, for example); followed by slower spikes signaling the object's spatiotemporal properties. The latter reflects the correlational structure of the inputs, including highly non-linear relationships. This information, typical of the structure of natural stimuli, is what is sent on to the visual cortex.[45]

Moreover, these dynamics are not fixed properties of LGN neurons. Feedback from the cortex, through those vast numbers of reciprocal connections, is now known to modulate the responses of LGN neurons to visual inputs. For example, Ian Andolina and colleagues manipulated feedback from cortex to LGN, and found marked changes in activity.[46] This suggests that the precision in stimulus-linked firing in the LGN appears as an emergent factor from the interactions between visual centres. It also suggests that the function of the LGN is to extract some information deeper than the 'what' and 'where' of static objects or features alone. But that abstraction only takes place with the help of networks 'higher up', a further stage of inference from the meeting of current with past experience.

Structure processing in the brain cortex

Similar findings have led to new views of what happens in the cerebral cortex, including the information it is most interested in, and of what happens to it. Neurons previously thought to be responding to discrete features appeared, on closer inspection, to be analysing more complex information as part of an extended neuronal ensemble. And it became clear that cortical cell responses are modifiable by context, experience and expectation, rather than constituting fixed feature-detectors.[47]

The idea that cortical neuron responses reflect a widescale 'petition', rather than a minor convergence, of inputs has attracted more support recently. Activity in one group of neurons is usually correlated with activity in many others, and the overall pattern of activity determines the responses of any individual cell within the group. As with studies on retina, these findings have reinforced the idea of population codes: the message is not in particular cells but in the activity of the ensemble.[48] To understand the signal from each neuron, the cortex has to listen to other neurons at the same time. This resolves as a 'coordinated coding' in that the signal is derived from the relations between multiple neurons in a population.[49]

But what is it they are saying to each other? What aspect of the signals or messages is the brain most interested in? Again focus is drawn to the deeper structural information in such messages. Much of this focus arose intuitively among some pre-war psychologists. As mentioned earlier, Gestalt psychologists proposed that the 'common fate' of moving optical patterns reflects intrinsically organised fields that are somehow directly detected by the brain.[50] In the 1980s, Donald Mackay argued that much of joint sensory behaviour (hearing while looking; moving eyes and head to sample a range of views; tactile exploration while seeing) has the purpose of collecting covariations (correlations), rather then preformed images. Even very simple exploratory probing of the environment by a moving fingertip exposes spatiotemporal covariation in the environment. He said that this explains why perfectly stable images projected on the retina (as mentioned earlier) quickly disappear: 'Stabilisation, even if it does not abolish all retinal signals, eliminates all covariation...If, then, seeing depends on the results of covariation analysis, there will be no seeing.'[51]

Of course, actually analysing the brain's use of deep structure is not easy, given the complexity of naturalistic stimuli, and the numbers of cells and synapses involved. But there have been some painstaking efforts. After noting that '[f]eatures associated with an object or its surfaces in natural scenes tend to vary coherently in space and time', Tai-Sing Lee and colleagues tried a multi-pronged approach. They first developed movies of 2D and 3D natural scenes and analysed the statistical structures in the data (including its fractal or deeper correlational structure). Then they implanted microelectrode arrays in experimental animals such as to record from hundreds of neurons in the visual cortex simultaneously. And then by exposing the animals to the movies they could assess whether the neurons actually use that structure.[52]

They showed, first, that the neurons do, indeed, fire preferentially to the deep spatiotemporal structure previously measured in the movies. Then they showed how such activities in the cortex tune neuron response properties and neuron connections. Finally, they suggest that reception of such structure in the cortex feeds back to 'enforce statistical constraints' on incoming stimuli – as, for example, those from the LGN. These constraints help extract the higher-order informational structure, eliminate alternative possibilities and resolve ambiguity during perceptual inference. Other studies have concurred with such findings of seamless interpretation of spatiotemporal data in cortical networks.[53]

These kinds of interaction between centres probably also explain the abundant reciprocal or recurrent connections between areas of cortex. Studies of rat, cat and monkey brains show that the average input–output connection ratio between different regions of the cortex is close to one. This indicates the extent to which each centre is engaged in cooperative information-processing with other centres. As mentioned above, the LGN, like other sensory way-stations, receives far more signals from the cortex than they send to it. This suggests the strong role of re-entrant (feedback) mechanisms, in which processing of inputs to cortical centres are rapidly used as feedback to 'firm up' information from their contributors.[54] This densely recurrent wiring organisation – especially the cortical–subcortical connectivity – has sometimes puzzled investigators. Susan Blackmore, for example, refers to the 'currently mysterious profusion of descending fibres in the visual system.'[55] But it is what is needed for definition of informational structures, as opposed to mere assembly of features or other unitary elements.

It is worth mentioning in passing that the significance of spatiotemporal structure has also been emerging in research in the auditory domain. In one well-known experiment it was first established that participants could easily distinguish between the sound of a bottle dropping onto a floor and bouncing (intact) and one in which the bottle shatters into bouncing fragments. The sound was then modified by putting its recorded acoustic spectrum through a sound synthesizer and damping out the frequencies corresponding with the explosive sounds of glass breaking.

It turned out that participants could still distinguish the modified sample from the one without the breaking glass. They did so merely from the way that the correlational structure of the acoustic samples unfolded over time. Listeners said things like 'that sounds like a bottle breaking', even though the acoustic spectrum corresponding with the breaking sounds, as such, had been removed.[56] Somewhat similarly, the study of rhythm, as in the psychology of music, is increasingly drawing attention to the deep covariation structure within it, or 'structure in time'. Again, we seem to appreciate such structure because of its tacit qualities of predictability, the essence of harmony.

In fact, direct recordings from auditory neurons confirm that they are highly sensitive to correlational structure in inputs.[57] But further clues come from the use of naturalistic auditory inputs. 'Neuronal responses in auditory cortex show...extremely complex context-dependent responses...(and) constantly adjust neuronal response properties to the statistics of the auditory scene. Evidence...suggests that the neuronal activity in primary auditory cortex represents sounds in terms of auditory objects rather than in terms of invariant acoustic features.'[58] More recent studies confirm that, as with vision, the acoustic structure of speech is captured in a non-linear pattern – a 'population' code – in a cell ensemble in the auditory system rather than independent neurons.[59]

Computations in synapses

That neuron networks in brain are interested in structures, rather than elements, has been confirmed from the way information is processed at the level of cell connections. Traditionally, it was thought that afferent (input) signals on dendrites are just summed at the cell body to generate the next signal (or spike train) from that cell, so long as a certain threshold has been exceeded. It is now clear that neurons can exercise computations that take into account the informational structure in inputs, probably through their interaction at the dendrites.[60] As Kenji Morita explains, with reference to specific cortical cells, 'there is cooperativity or associativity between nearby synapses in the dendritic tree of individual cortical cells'.[61] Such processing hugely amplifies computational and storage power, because a relatively few interaction parameters (e.g., correlations and multiple correlations) can be used to generate unknown values – more complete patterns – from incomplete current inputs.

To suggest a crude example, the covariation between two variables A and B (say the speed of motion of two light spots) might be expressed as A=2B. Once that parameter has been registered in a neural network, then B can be predicted given any value for A in input without the necessity of having to store the point to point associations:

(A=1)→(B=2)
(A=2)→(B=4)
(A=3)→(B=6),

and so on. More realistically the covariation will be conditioned by a third and additional variables, as in A=(2B)xC, and so on. Finally, registration of interaction parameters permits non-linear processing, with attendant increase in creativity of response.

This kind of information storage and computational process has recently been traced, in fact, at the complex biochemical level within the post-synaptic junctions.[62] When we consider that each cortical neuron may receive inputs from 100,000 other neurons, and send impulses to 100,000 others, then we can begin to see that the computational capacity of brain networks is huge. This is complexity indeed.

The phenomenology of sensory structure

Donald Mackay was one of the first to stress that the brain's extraction of spatiotemporal structure from experience makes rich predictability available, even from meagre data. For example, a blind person can discern the pattern on a manhole cover by probing it with a cane. Every time we park a motorcar we can sense the position of wheels and bumpers, even though we cannot actually see them.

In fact we take this structure in experience so much for granted that we scarcely notice it. It only tends to become apparent in certain situations when it is either deficient or overwhelming, in some way. This may explain the common feeling of bewilderment, or losing touch with reality, as in a mountain white-out – or in busy shopping malls! But it is seen in perceptual systems that have been rendered defective in some way. For example, Hermann von Helmholtz described (in 1909) the visual experiences of patients who had gained sight for the first time after operations from congenital cataracts. One patient was surprised that a coin, which is round, should so drastically change its shape when it is rotated (becoming elliptical in projection). Another found it seemingly impossible that a picture of his father's face, the size of which he knew from touch, could fit into a small locket.[63]

Environmental complexity, brain and evolution

Environmental complexity, or spatiotemporal structure, as the processing currency of the brain, suggests a consistent evolutionary story. The processes that evolved within cells for assimilating environmental structure, using receptor and internal signaling dynamics, have, in nervous systems, evolved to process the structure of traffic flow between cells in vast networks. Most significant of this nesting of pre-evolved processes into new functions is that of developmental (including gene regulation/epigenetic) processes to furnish lifelong brain plasticity and learning. Obviously, the more extensive the network, the deeper the informational structure that can be processed. The intricate connectivity of brain networks has increased enormously in the evolution of mammals. This has been achieved largely by increased folding of cerebral cortex to obtain greater surface area and, therefore, volume for connectivity. Yet the cerebral cortex consists of a remarkably uniform structure consisting of six layers of cells in a sheet no thicker than a credit card, but with abundant interconnections within and between specific areas of cortex and subcortex. The theme seems to be one of expanding processing capacity for abstraction of ever-deeper spatiotemporal structure, matching that increasingly experienced in more demanding environments.

This relationship is evident in brain network development in several ways. It is seen conspicuously in the increasing developmental plasticity in brain (and, with it, of behaviour and cognition) over evolutionary time. It has long been known that connectivities of regions of the cerebral cortex are largely experience-dependent (see Chapter 5). Also, circuits in even the simplest brains in nematode worms and insects are capable of learning and memory. Although members of a species have the same complement of initial connections in the embryo or foetus, the density of specific cortical fibre pathways can vary substantially between mature brains.[64] In humans, this variation often reflects occupational specialisation. London taxi drivers, who are required to develop a detailed memory of street layouts, have part of the brain involved in spatial memory (the posterior hippocampus) enlarged.[65] Likewise, the part of the cortex involved in the sensorimotor aspects of finger coordination in violin players is expanded on the corresponding side of the brain, but not on the other.[66]

The interesting thing is the dependence of such development, not merely on environmental stimulation, but also on environmental *structure* in experience. As mentioned in Chapter 5, spatiotemporally structured light, rather than just any light stimulation, appears to be required for proper development of visual cortex. Confining visual experience to white noise, that presents all light frequencies without the patterned input, retards development of connectivity.[67] Indeed, just before the onset of sensory experience, brain networks seem to prepare themselves for the reception of patterned activity. For example, retinal ganglion cells fire bursts of outputs that are highly correlated temporally and spatially, as if setting up the system for sensitivity to the structure in eventual experience.

Experiments on dark rearing in ferrets suggest that the response properties of cortical neurons is rudimentary without the structured experience needed for full tuning. For example, lid-sutured individuals (an approach to light-exclusion in development) suffered a more devastating effect on development than even those reared in total darkness (without suturing). The investigators speculate that it is the abnormal scattering of light experienced under lid-suturing that causes the problem.[68] Similarly, random sound or noise is not sufficient for proper development of auditory cortex.[69] It all suggests that the cerebral cortex, with its extraordinary plasticity, seems to have co-evolved with complex environments for the capture of environmental structure.

Brain dynamics

Not surprisingly, many investigators now argue that both the development and ongoing functioning of brain can only be understood within a dynamic systems framework. There is ample scope for limit-cycle and chaotic dynamics in the rich connectivity, non-linearity of traffic and abundant feedback signaling, in brain networks. Even in fruitflies, studies of the processing of light stimuli show how responses in certain cells exhibit second-order, non-linear interactions among motion detectors.[70]

The advantage of such dynamics is that they permit rapid resolution of the novel sets of inputs with which the system is being constantly bombarded. 'The most significant property of ensembles is the capacity for undergoing rapid and repeated global state changes. Examples are the abrupt reorganizations manifested in the patterns of neural activity in the brain and spinal cord by the transitions between walking and running, speaking and swallowing, sleeping and waking, and more generally the staccato flow of thoughts and mental images.'[71] As Freeman explains, brains are intrinsically unstable and, in effect, continually create themselves.

More importantly, such dynamics, complex as they are, take us from neural to cognitive activity, and to much greater complexity. This is the subject of the next chapter. In the meantime, I hope you can see how the nature of brain networks, and their attunements to environmental structure, provides yet another important bridge to the evolution of greater complexity in organisms.

7
From Neurons to Cognition

A major problem in understanding mental functions has been: not grasping how to get from the intercommunication between neurons, in their vast networks, to cognitive functions, which seem to be on a different plane, expressing a completely different *kind* of function. How do we relate the two levels? This difficulty is, I think, the main barrier to understanding cognition as a complex evolved system. It is reflected in a kind of mind-brain or mind-body dualism in theories, creating a number of persistent problems in cognitive sciences, as well as a lack of conviction and coherence.

One of the problems is how stable, symbol-like percepts or images can be detected in fleeting, ever-changing inputs, as we have just seen. Another is the 'binding' problem, or how different aspects of experience, supposedly detected as separate 'pieces', can then be put back together in the brain as a coherent image. Yet another is the 'grounding' problem, or how inputs, allegedly processed as distinct symbols, ever derive their 'meaning' and utility for action. This implies another puzzle, which is the 'meaning of meaning' itself. These problems have dogged cognitive psychology from way back.

We have already seen how extensive nerve networks, with complex signaling and epigenetic systems, and lifelong developmental plasticity, evolved. In these networks, signals generated from neurons are continually modified by the dynamics of signaling *between* them. Cognition obviously somehow resides in that rich and intensive chatter between neurons. In this chapter, I want to describe how the most basic cognitive processes of recognition and classification emerge from that chatter. I will turn in the next chapter to what are usually considered to be the 'higher' cognitive processes of learning, thinking, decision-making, motor action and so on.

The emergence of cognition

Cognitive systems of sorts arrived early in the course of evolution. But they became particularly strongly demanded when objects became the main sources of environmental complexity. In single-cell organisms, molecular forces and epigenetic and other signalling systems do a good job in dealing with general and physical changes, as we have seen. As animals themselves became bigger and more motile, though, their worlds became more 'objectified': visual objects, auditory objects, tactile objects and even olfactory objects. After the upsurge and diversification of animal life about 500mya (the Cambrian explosion), other animals probably became the most significant objects to one and all. Chasing, or being chased by, another, means constructing a rapidly changing perception of that other within the context of an already changing inanimate environment. Trying to make sense of such objects, as opposed to physical or chemical gradients, or static barriers, involves sensory signals changing in far more rapid and complex ways. The animal constantly needs to predict things such as what comes next, the best next move and its probable consequences.

More specifically, the animal needs the ability to integrate these moving, transforming, fleeting, and usually incomplete, impressions into some more definite internal 'image' of the object. It needs the ability to predict the behaviour of the object, either by itself, or, more usually, in combination with that of numerous other objects simultaneously. And it needs to be able to fashion fruitful responses, even while the scene is changing, and being changed by its own activity. These are usually what we mean by *cognitive* functions.

The unique function of cognitive systems, then, is to disambiguate confusing, incomplete and rapidly changing data, while generating equally unique, yet adaptive, responses. As mentioned in the previous chapter, this basic problem can only be overcome through some kind of inferential or inductive processes within the system itself, making predictions from stored structural parameters. This basic inferential-creative function evolved very early, and is found even in the miniature brains of insects.[1] The brain of the honeybee, for example, has less than a million neurons, yet is capable of cognitive functions.[2] In their foraging outside the hive, bees navigate large areas using spatiotemporal relations between landmarks, sun position and patterns of polarised light reflected from the sky. They also visit hundreds of flowers in gathering food, inspect potential new nest sites, learn and memorise their locations and communicate to others through a language called the 'waggle dance'. They learn categorisations based on generalised properties such as 'vertical' and 'horizontal', still a basic cognitive function in advanced systems, as we will see. And they must productively engage with the seething, yet somehow orderly, social activities in the hive.

There are other powers of abstraction, too. After foraging trips, bees don't find their way home by simply retracing their outward meanderings: they memorise and integrate the directions and distances experienced, along with motion cues, to construct the most direct route, as if having acquired statistical structural parameters. For example, in one study, bees were trained to forage at two feeding sites, one only in the morning at 630 metres from the hive and 115° north, the other only in the afternoon at 790 metres from the hive and 40° north. The bees were subsequently released at the 'wrong' site, either the afternoon site in the morning, or *vice versa*. They nevertheless flew back to the hive in a direct line.

According to Lin Chen and his colleagues, this kind of ability suggests that the bees have internalised the 'higher order' relational invariants from prior experience of the layout of the world.[3] Rather than simple cue-response functions, this learning reflects 'a dynamic and self-organizing process of information storage', as Lehrer and colleagues put it.[4] Other research over the last few years has revealed previously unsuspected complexity of cognitive processing in invertebrates generally.[5] Fast forward half a billion years of evolution and ask, what, then, of the cognitive systems of humans?

Theoretical dissonance

In fact, there are many disagreements about the nature of cognitive functions – what a cognitive system is for – because investigators are still not clear about why and how cognitive systems have evolved and become more complex. Arguments about the nature of cognition have raged since Ancient Greece and beyond. The literature abounds with the diversity of models that still exist, and their varied assumptions and histories.[6] Since the 1960s, however, cognitive theory has been dominated by a fairly standard computational model (SCM). This model originated with the inventor of the computer, Alan Turing, who defined human mentality itself (thought, knowledge and intelligence) as systems of computations.[7] He said that electronic computers would eventually come to exceed the cognitive powers of humans, as well as other animals. The basic idea is that of responses to input 'stimuli' according to series of rules that transform the stimuli and send them on to the next set of rules, and so on, until a response or output is created. On today's digital computers, the stimuli are usually input at a keyboard, and the rules are in the form of programs or algorithms stored on silicon chips. There is a memory for past impressions, such as associations between stimuli, or between stimuli and past responses. And a 'central executive' acts as overall manager of processing.

From the 1950s onwards, the model formed the basis of 'artificial intelligence', and inspired much related theory in animal and human cognition. I cannot enter into a detailed critique, here; so suffice to say that artificial intelligence has produced machines that do simple, routine things (such as select and count items according to fixed criteria) very well, but cannot deal with fuzzy, dynamic inputs such as moving images or human speech. In his book, *The Mind Doesn't Work That Way* (a riposte to works like Pinker's *How the Mind Works*), Jerry Fodor admits that such attempts have failed to produce successful simulations of 'routine commonsense cognitive competences.'[8] He reminds us that no one has managed to show how sensory data come nicely carved up in the right way, as in discrete characters at a computer keyboard, for it to be appropriately channelled into the kind of computer programs conceived.

An alternative approach has used computers in a different way. Investigators set up model neurons (within software programmes) with inputs from the 'environment' (in fact the keyboard). The input neurons are connected to other model neurons in series of layers, all obeying certain rules. If the sum of the inputs on a unit exceed a certain threshold, then that unit transmits a signal to whichever other units it is connected to – and so on, through the different layers. The overall 'wiring' in these miniature networks is such as to allow associations to form between the activities of the units according to associations in inputs. For example, a pair of units will tend to respond together if the stimuli, which they were set up to respond to, also tend to occur together over a sequence of inputs. That happens because the connections can be 'weighted' (strengthened or weakened) by part of the programme according to frequency of use. The response properties of the units are duly modified in the process, just like the real thing. In that way they can record, in their 'synaptic weights', something like the statistical structure of the 'environment'. Figure 7.1 shows the idea in very simple form, although I stress that many more complex kinds of networks are used.

Simple as they seem, experiments with such artificial neural networks (ANNs), as they have been called, have given us some new insights into

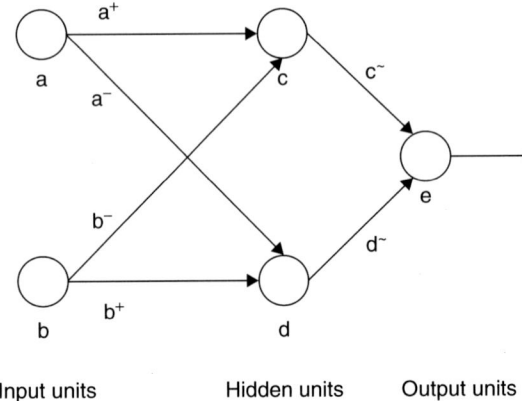

Figure 7.1 Fragment of an artificial neural network. Stimuli (usually some well-defined features, as conceived by the experimenter) are input at the keyboard and received by the input units. If the strength of signal exceeds a certain threshold, the unit discharges along its connections to other units. The connections can be excitatory or inhibitory (e.g. a+/a-). After integration of a series of signals in subsequent units, the outputs should reflect statistical associations between the inputs

cognition. Experimenters have been excited by the way that complex patterns of weights can emerge that approximate complex, abstract rules, as in categorisation and simple language learning.[9] That is, they may exhibit 'emergent properties' like those typically found in cognitive systems, reflecting the deeper structure in experience without anything very much being 'built in'. Earlier networks were criticised, literally for their artificiality, including training being guaranteed by the feedback from the operator, the use of unreal digitised inputs and so on. However, 'self-supervised' networks, which acquire structure, and generate feedback, automatically, are now in common use. More recently it has been discovered that non-linear/chaotic dynamics can occur within ANNs, and it is such findings I will be referring to later.[10]

Dynamic cognition

It is in this context, coupled with findings from brain research, with developments in the mathematics of non-linear dynamics, and the availability of high speed, high capacity computers, that dynamical views of cognition have become very popular over the last few years. In a recent comment, Ulric Neisser, one of the founders of the earlier computational model, says that the dynamic approach puts cognitive science itself on a new trajectory'.[11] Within the general approach, however, there are various emphases.

Some researchers have simply stressed the temporal continuity of experience and its implications for cognitive functions. (Again, it seems highly surprising, that this is only a recent discovery). Michael Spivey and Rick Dale say that, 'Rather than a sequence of logical operations performed on discrete symbols, real-time cognition is better described as continuously changing patterns of neuronal activity. The continuity in these dynamics indicates that, in between describable states of mind, much of our mental activity does not lend itself to the linguistic labels relied on by much of psychology'.[12]

Spivey and colleagues have adopted a widely used research strategy. This is to focus on a particular aspect of cognition, gather data from tasks thought to tap it, describe how the data can be accounted for by dynamic processes and then perform computer simulations to support that. For example, they presented participants with on-screen problems on a computer and recorded mouse-movements and eye movements as they followed decisions emerging over time. They claim that the sequences reflect patterns of neural activation passing through brain states and falling in to, or towards, attractor basins. The latter are thought to correspond with specific perceptions, conceptions or possible actions.

Other researchers have stressed the importance of self-organised criticality, including chaotic dynamics, in cognition, and searched for evidence for it. This view proposes brain/cognition in constant non-equilibrium states, close to criticality. In hearing a sound, detecting an odour, scanning for predator or prey, a speedy search of the environment is essential, as is a speedy, novel, response. Such brain states exhibit maximum sensitivity to inputs and are open to rapid state transitions from chaos to order and back again. Foremost among proponents of this view has been Walter Freeman, who notes how '[t]he process is by neurodynamics, not by logical rule-driven symbol manipulation.'[13] Over a period of forty years or more, Freeman and colleagues have studied those dynamics through the neuroelectrical signals seen in elecroencephalography (EEG, popularly known as brain waves) recorded from the scalps of humans and animals.

Freeman argues that brains are essentially non-equilibrium systems, virtually never in steady states due to constant perturbation. Brain networks use dynamic patterns of activation to execute cognitive tasks. The non-linear properties and criticality of dynamic processes are key properties. They foster robust operation, and fast and reliable decision-making, in the face of changing environments. Brain activity, agitated by constant inputs, moves constantly over an attractor landscape. Those attractors entered by a stimulus respond by temporarily assuming critical or chaotic states until the stimulus is constrained to a meaningful identity (output pattern). The advantages of dynamical processing in cognition include greater processing capacity, speed of processing, creativity, stability and robustness in the face of uncertain inputs.[14]

Evidence for chaotic activity in the brain has been gathered from several sources. Much of it comes from the EEG studies of Freeman and others. As mentioned in Chapter 6, single-neuron and network firing patterns also exhibit dynamic features of chaotic attractors.[15] John Beggs and his colleagues have grown thousands of interconnected neurons on a mesh of electrodes. These 'brains in a dish' can be kept alive for weeks while their spontaneous electrical activity is recorded. The abundant data indicate metastable states on the edge of criticality.[16]

Other evidence of 'deep structure' in cognitive processes further supports the idea. Remember from Chapter 2, how well-structured phenomena, throughout nature, tend to have a fractal structure. This means that they are self-similar at many different levels: as with the tree and coastline examples, there is structure both within and across levels that impart predictability (shown the extremities of a tree branch partially obscured you should be able to fill-in the missing sub-branches). But dynamic processes, changing over time and space, can also be fractal, equally reflecting deeper structure. There is now abundant evidence that cognitive processing has such a fractal (nested) structure.[17]

Other evidence for non-linear dynamics in cognition lies in observable phase changes in human cognition. Damien Stephen and colleagues detected self-organised cognitive dynamics in the sudden emergence of new insights in learning and problem-solving. Young children commonly add 4+2 by sticking up four fingers on one hand and two on the other, and then counting them through. Later they simply supplement from 4, then 5 to reach 6, in succession. As Stephen and colleagues say, 'the change in structure appears to be driven by the activity of the system itself. There is no external agent guiding the individual toward a new organization, nor is there an internal plan that contains the new structure in miniature.'[18]

In another study they asked students to solve a problem involving a series of gears. The solution required the discovery of a particular relationship between the gears. The researchers followed the internal cognitive dynamics by monitoring changes in eye movements over time, using two measures. The first, entropy, is a measure of the organisation/disorganisation in a system (as mentioned in Chapter 2). The second was a measure of the degree of 'nestedness', or deeper structure, in the eye movements over time. Both measures were consistent with change from one attractor structure to another due to the dynamics of self-organisation within the system.

Finally, these conclusions have been strongly supported by recent work on artificial neural networks. More recent ANNs have gone well beyond simple associations to mimic real brain dynamics more closely. Of particular value have been ANNs using recurrent (feedback) connections from the output layers to the input layers. These use the history of network experience to adjust, and narrow the variability of experience, much as cortical feedback connections do in the brain. Moreover, as the correlational structure builds up in the network, so the spontaneous dynamics of the ANN becomes increasingly chaotic, and yet increasingly structured by structured inputs.[19] The implication is, the more you learn the more likely the reaction is to be chaotic, and the greater the precision of processing.[20]

Now let us see how these self-organising principles might work in creating cognitive states from the mere firing between neurons – and thus to clarify the distinction between the two.

Dynamics take us from neurons to cognition

The distinctive demand on cognitive systems is that they need, first, to be very good at assimilating structures that change frequently – more quickly than can be dealt with by epigenetic or developmental systems; and, second, using that structure to predict immediate and distant futures from inevitably partial current information. As seen in the previous chapter, the brain is superbly evolved for supporting just such functions. The responses of single neurons to sensory stimuli are highly variable and 'noisy'. Yet our *cognitive* experience of the environment is much more stable and consistent.

So how do we get from noisy brain processes to consistent cognitive functions? In the rest of this chapter, I want to offer a purely non-technical illustration of how this might happen. I will be content to show this, first, with regard to very basic cognitive operations in a generalised animal cognitive system, and thus survival-based action in a highly changeable world. In the next chapter, I will extend this to more complex cognitive functions. (In the final chapters, I will show that something else very significant happened in the course of evolution of the human cognitive system).

We have a framework for the creation of cognitive states from brain states in Walter Freeman's studies of EEGs during olfaction in mice. Each smell produces 'a global wave of activity' in the first subcortical level, the olfactory bulb. This activity is rapidly transmitted to other cortical and subcortical centres. Feedback information from those centres, in turn, creates chaotic activity back in the olfactory bulb. This chaotic activity quickly resolves into a brief limit-cycle, or more stable state, representing the smell it is now perceiving. Finally, Freeman notes, each such attractor rapidly emerges and collapses with each inhalation-exhalation 'so that the system is freed to make another test.'[21] Freeman cites evidence that similar kinds of non-linear dynamics hold for visual, auditory, and somatosensory, as well as olfactory inputs.

Freeman and colleagues suggest, then, that it is not external smells *per se* that animals respond to, at least directly. Rather they respond to internal activity patterns created by the chaotic dynamics within the olfactory bulb. The non-linear, chaotic dynamics of the system seem to be an essential part of the process. This is because 'neither linear operations nor point and limit cycle attractors can create novel patterns... [T]he perceptual message that is sent on into the forebrain is a construction, not the residue of a filter or a computational algorithm.'[22] It is this construction, or activity pattern, that is now a cognitive agent, rather than a mere neural one, because it now enters into a new level of regulations with other such agents, creating new properties of life in the process.

Abstraction of a primary structure

The basic primitives of visual experience, are usually described as 'features', even though scientists disagree about what that actually means. But even these are abstractions that turn neural network activities into cognitive experience. Visual experience starts as objects traverse the visual field (and/or as we move around them, handle them and so on). This experience consists, in effect, of a rapidly changing mosaic of light energy on the retina: a 'cloud' of moving points exciting areas of photoreceptors in the retina in seemingly disorderly succession (see Chapter 6).

But the cloud isn't really disorderly. By the nature of the structure of the objects it emanates from, this changing 2D mosaic of light on the retina is not a random pattern. Rather, it reflects the fact that everyday, moment-to-moment experience tends to present itself as a whole global structure, coherent over space and time. Accordingly, statistical analyses of natural stimuli, mentioned in Chapter 6, have revealed numerous correlations between points at various statistical levels or depths. That is, many aspects of the 'light spots' on the retina – their changing intensities, directions and speeds – will covary over space and time.

A cluster of light spots on a corner of a book, for example, will stand out, in the way that they move together over time. The form of that covariation makes it distinct from other clusters on sides and edges, and from numerous other points that are uncorrelated. It is such spatiotemporal structures that are rapidly picked up by retinal networks evolved to be sensitive to them. But in order to become *cognitive* structures, information needs to be added, or 'filled-in'. This comes from the information within the connection patterns of the networks: information derived from previous experience with similar objects and consisting of the relational parameters making up the structure within them. These attunements are the attractors to which incoming stimuli bearing similar structure are pulled (Figure 7.2).

Evidence for this kind of proto-featural construction is abundant. Much of it was mentioned in the context of cell responses in brain regions in Chapter 6. Figure 7.3 shows perhaps the simplest structural feature of objects: an edge of an object, or 'line', as an array of light points falling on the retina. In reality this would consist of thousands of light points, of course, moving rapidly across the visual field. But only three of them are shown here, as if in the briefest of snapshots; and they are much enlarged for present purposes.

This particular sequence, in fact, is extracted from a video of the movements of the edge of a chair as viewed by someone passing around

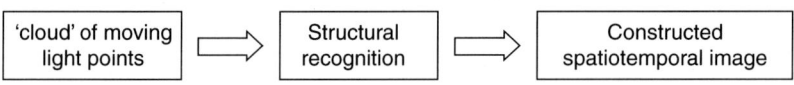

Figure 7.2 Extraction of covariation structure in early processing

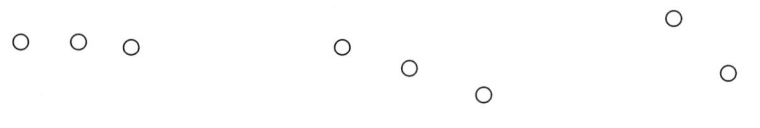

Figure 7.3 Experiencing a 'line'

and towards it. As can be seen, there is (superficially, at least) no overt feature, at all, only a sequence of light points. Shown as a static entity, that is, the trio is not very informative. Show the sequence on a computer screen, though, at its natural speed, and people immediately report seeing a 'line' looming from the distance, or even the sensation of movement towards and around it. That is, the dynamic, spatiotemporal form evokes a compelling picture in which people interpolate the large number of missing points, and their relevant positions, even with this most sparse of inputs.

There is no computer that can deal with such sparse, fuzzy – and probably entirely novel – visual data like these and create a cognitive construct from it, especially so rapidly and reliably. So something is obviously being added by the system; probably something abstracted from experience with similar features in the past. As expected, analyses of the statistical relationships between even these three points, as they move in space and time, indicated considerable structure between them. Extended to the dozens of light points in the original complete line, this simple example suggests a very rich source of information from which invariant structural parameters, typical of many in experience, can be extracted. Obviously, that structure will often include changes in other variables, such as intensity and colour, not just the changes in position. The complete experience will also include other dynamics of the feature (or object) as a whole, such as rotation, looming (dilation) and spiralling.

We all have an overwhelming tendency to think of an experienced 'line' like this, somehow being taken into, and circulating around, the brain as a 3D iconic image, as if projected from a cine camera onto an internal screen. Yet it's obvious, that couldn't possibly be the case. On the other hand, we know that each experience can be uniquely specified by its spatiotemporal (structural) parameters. These may be simple bi-variant correlations, or more complex associations at several depths. For example, the rate of change of two light 'spots' may be associated (that's one parameter). That association may itself be associated with the rate of change of another light spot (that's another parameter); and so on, across the myriad light spots and their numerous depths of association. Obviously, there may be dozens or hundreds of such parameter values available to specify the specific experience. And we know that neurons are sensitive to such associations. So, it seems likely that this is the form – distinctive sets of parameter values – in which even simple 'features' like that are internalised and passed around in the brain.

Of course, increasing experience with features like that will reveal increasing numbers of structural parameters, and these, as registered in network connections, will accumulate as a basin of attraction for that feature, with the deeper levels representing increased sharing of parameters: the deeper the levels the closer a set of parameters is to the prototypical 'line' (Figure 7.4). The current, fuzzy, experience, itself full of 'sample' structural parameters, can then be drawn to the attractor(s) with corresponding structural parameters. Note that the attractor itself is not in a stable, equilibrium state, with one dominant firing pattern. The current stimulus, as well as top-down inputs from other centres (which contain crucial contextual information), put the attractor into a chaotic state. The parameters detected in the (sparse) inputs activate those stored from previous experience. The chaotic attractor trajects around all permutations of variable values, including 'missing' light spots, consistent with those parameters, until the 'best' expression – a more complete line – is obtained. The accessed

Figure 7.4 A new stimulus (arrows), moving as a wave of activation containing 'sample' feature parameters, is drawn towards basins of attraction containing corresponding sets of parameters ('contour' lines, or isoclines, connect parameters that have occurred with similar frequencies in previous experience, those at deeper levels representing the more coherent sets, e.g. the more 'definite' line image)

correlational parameters serve as a kind of 'grammar' from which a coherent 'sentence' is obtained from a fuzzy piece of 'speech'. That pattern of activity is discharged to other centres. This is the internal 'image' of the line in current experience.

That output is now a 'cognitive' agent – a percept – rather than just a neural one, even at this very early stage of sensory processing. The perception of even a simple image like this consists of a continuously constructive process, changing even while we're looking at it. We really have to stop thinking in terms of iconic images somehow being taken in 'whole' and passed around in neural network space. As an emergent cognitive construction, the image now enters into a new level of activity, with other similar constructs, with new rules, being able to predict so much more than before. At the same time, the network itself becomes reconfigured (connection strengths between synapses are adjusted) to be even better prepared for the future.

Of course, other experiences, such as a moving curve with a few additional points, will exhibit a *different* covariation structure, tending to find a different attractor in the network, and evoke different (neural and cognitive) outputs. There will be attractors corresponding with the hundreds of local 'clusters' of structure at that primary level. In all cases, though, covariation structures are being detected in sensory inputs. They are pulled towards corresponding attractors where reciprocal interactions will come into play to predict, and fill-in, most likely missing details. This dynamic generation of images from stored parameters contrasts with the implausible theory that the (constantly novel) images are just picked up from sensation through a process of feature-detection.

Construction of more complex images

In the interactions within and between networks, abstraction of input structure, and construction of meaningful cognitive images, continues, so long as such a structure continues to be found. The outputs from the myriad initial attractors then become integrated at higher levels because of the spatiotemporal parameters their outputs will tend to share. So far we have created a percept, or cognitive image, of a line. But just as the simplest features (lines, edges) are characterised by their deep spatiotemporal structure, so these structures themselves, as aspects of the same current experience, will tend to covary with one another, in analogous ways, at still deeper levels. Structural parameters corresponding with that deeper level will also have been abstracted in previous experience. The dynamic interactions just described seamlessly continue.[23]

For example, the covariation patterns that define sides and edges of the book you were looking at will themselves exhibit covariation at a deeper level. As you move around it, *they* move together and exhibit an inter-featural structure peculiar to that object or class of objects. Likewise, as we move around a chair, the line cluster that is picked up from the front will covary in space and time with the line cluster emerging from the edge or back of the seat. This joint pattern reflects a deeper covariation structure that can also be registered by brain networks sensitive to such structures, and within which the simpler structures are statistically nested. This will potentially create further basins of attraction at a higher network level. Again, through network interaction and chaotic activity, novel cognitive images of more compound features can be constructed.

Evidence for the dynamic generation of complex cognitive images from nested covariation structures comes from experiments with point light stimuli (PLS). In these studies, small patches of reflective material are attached to prominent aspects of objects. The objects are then filmed in the dark with a light shining on them as they move under natural, dynamic circumstances, or as they might be seen by someone walking around them. The object thus appears only as a small set of light points moving in a more or less coordinated manner. The best example is the point light walker, as shown in Chapter 6 (Figure 6.1). Presentation to observers as a set of static points evokes little recognition of the source object: although there is neural activity, there is literally no cognition. Within a second or two of movement commencing, however, the underlying structure becomes apparent and observers accurately label the source as a person walking.

Application of statistical models to samples of point light walkers have revealed considerable covariation structure within them, at various statistical levels. Some of these parameters are simple correlations. But there are also many first-order interactions, second-order interactions and so on. We attempted to show the role of such structure by stripping out one or more points in turn and checking how much structure was left. In this way, in one experiment, we arrived at ten different permutations of four points each, selected in order to display different degrees of covariation complexity (measured through statistical models). All else was equal. Sequences of these very sparse PLS were then presented to participants, who were asked if they recognised any objects. (They were interspersed among sequences of 'nonsense' PLS as distractors).

Across the range of stimuli, there was a significant association between the covariation complexity in the PLS and the numbers of observers who recognised a person walking in the display.[24]

This rapidity and accuracy of response suggests that recognition from point light stimuli is a further spectacular example of image construction from sparse inputs, based on the dynamic interactions. The system has abstracted the dynamic structural parameters of people walking, from previous experience. The current structure (the PLS) is drawn into the corresponding attractor; which then discharges with more coherent, meaningful, output; but now as cognitive objects, not just neural ones.

This isn't magic, conjuring information out of thin air, or getting something for nothing. It is part of the nature of complex information. Every time you infer an object or event from a fragment of stimulus, this complementation is what you are doing. This utilisation of deep covariation structure in changeable environments is precisely what the cognitive system evolved for.

Construction of object images and concepts

There will almost certainly be much deeper structure in experience than that just described. A system evolved to capture it will tend to do just that, so long as it has the capacity to do so. Accordingly, the nested feature-structures just described will integrate at a still deeper level, corresponding with the cognitive images of whole objects. Again, this follows from their shared spatiotemporal structure. As an object is experienced in everyday perspective its features and parts flow and change, not in random isolation, but with spatiotemporal covariation. The seat, back and legs of a chair move together in the visual field in a coordinated manner. Those spatiotemporal parameters are duly registered in brain networks. They are then used to dynamically construct the internal experience or image of that object from the current inputs.

The fundamental cognitive phenomenon of a concept forms from the structural parameters shared between individual objects in a particular category. While some of that dynamic structure, and sets of parameters, will be peculiar to individual objects, such as buttercups in a field, or roses in the hedgerow, most of it is nevertheless shared with objects of the same category. To a bee they are all flowers. There may be many different chairs, but they are all chairs. Over experience, the shared structure and predictabilities will be self-organised in networks as attractors corresponding with those particular categories. The attractor is a cognitive structure called a concept. The shared structure within that category will, of course, be radically different from that in other categories such as leaves or stems. Different levels within the attractor will reflect degrees of shared structural parameters, as indicated in the 'contours' in Figure 7.4. The more a sample input, such as a specific object, combines parameters from those deeper levels, the more readily it will tend to be recognised as close to a 'protypical' example of that category, whether or not it has actually been experienced before. This has been well demonstrated in research. So we have the basis for a whole system of concepts, which also come to serve as powerful cognitive agents. The basic process has been well demonstrated in ANNs, and is implicit even in insect brains, as mentioned earlier.

That structural basis of concepts was demonstrated in some other studies with point light stimuli. These were designed as if from moving animals (sheep walking; dog barking and so on). Although the stimuli were very sparse – as few as three or four points – even 8-year old children were able to distinguish between the different categories on the basis of the structural parameters within them.[25] This also suggests a basis of cognitive development. As experience in a domain increases (or as children get older), brain networks quickly become attuned to increasingly extensive and deeper structural parameters in a domain.[26] The same idea, of course, explains differences between novices and experts in many domains.

Such complementation, or filling-in, from sparse inputs has been much discussed in recent years: many studies show that what we

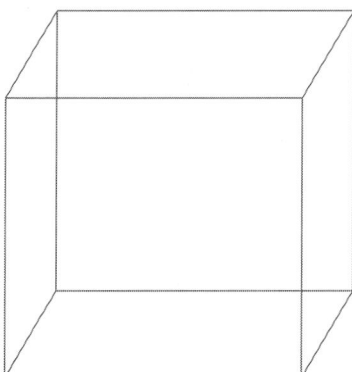

Figure 7.5 The Necker cube

perceive can be different from what we sense. The phenomenon of illusory contours, as with the Kanizsa triangle, was mentioned in the previous chapter. Not only do we add lines to this figure where they are predicted to be, but neurons in the visual cortex fire as if they really are present. The fluidity of the dynamics may also explain the possibility of two or more rival attractors within an attuned network. For example, the Necker cube illusion appears as an image that flip-flops between two or more equally suitable possibilities, or attractors (Figure 7.5).

Richard Gregory has argued over many years that visual illusions arise from the restructuring of current experience based on stored knowledge. He has shown, for example, that patients recovering from lifetime blindness are not susceptible to typical visual illusions.[27] Presumably, this is because they have not acquired the deep structural parameters, and developed corresponding attractors, that occasionally generate the illusory outputs in the rest of us.

The main function of concepts/categories as dynamic cognitive agents is that, again, in rapidly changing environments, they furnish enormous predictability very quickly. Once an animal has recognised an object as belonging to a particular category (drawn the input activity into a structurally congenial attractor basin) it can predict much more about it – including how to respond to it. Research on categorisation and concepts has made us acutely aware of the astonishing abilities of this system of predictability. 'The visual system has an extraordinary capability to extract categorical information from complex natural scenes. For example, subjects are able to rapidly detect the presence of object categories such as animals or vehicles in new scenes that are presented very briefly... This is even true when subjects do not pay attention to the scenes and simultaneously perform an unrelated attentionally demanding task... a stark contrast to the capacity limitations predicted by most theories of visual attention.'[28]

Marius Peelen and colleagues say that 'the rapid detection of categorical information in natural scenes is mediated by a category-specific biasing mechanism'.[29] We now have a better idea of what that biasing mechanism consists of, and why it is so extraordinarily productive. In that productiveness, though, lies the solution to another puzzle, earlier mentioned as the 'meaning of meaning'. In this dynamical view, the meaning of an object image lies in all that it predicts. In acquiring that predictability, and enshrining such connectedness with the outside world, an image becomes a 'symbol', pregnant with predictive possibilities. This also explains a symbol's 'grounding' in reality: its whole semantic content consists of that connectedness with the world of experience. This is why apprehension of 'meaning' increases with brain evolution.[30]

Here, again, we are describing the emergence of a new level of activity and potency, with its own laws and principles, based on vastly enhanced levels of predictability. The principles are properties of neural activity, sure enough. But a new level of regulation has emerged that utilises and commands neural space rather than vice versa.

Multi-sensory cognitive images

We have so far been considering the emergence of cognitive functions in vision, together with some excellent work on olfaction. But natural experience is a multi-sensory phenomenon, and this is being increasingly appreciated in research and theory.[31] Attractor landscapes evolve in much the same way in networks in other sense modes. Perceptual experiments have shown that the learning of deep statistical structure occurs in auditory and tactile sensory channels, as well as vision.[32] I mentioned in the previous chapter that 'structured' sound is necessary for proper development of the auditory system, not just random noise. Israel Nelken reviews much evidence for the on-line extraction of statistical regularities from the auditory scene, also suggesting that the organisation of the auditory scene manifests in terms of auditory 'objects' – analogues to constructed visual objects – rather than in terms of isolated acoustic features.[33]

There will be considerable covariation between the statistical structures sensed in different modes: touch with vision, vision with sound and so on. This is particularly the case as active movement and action on objects will change experience in several sense modes simultaneously: vision and sound of a person walking, for example. There are also signals from internal senses. Every time an individual acts on objects the brain also receives a barrage of signals from receptors within the body itself: stretch receptors in muscles and tendons; various kinds of touch receptors in skin; pressure receptors in joints and muscles; and from the muscle spindles (providing information about body posture). Changes in all of these will covary with those occurring simultaneously in visual and auditory experience of the event.

As with previous levels of perception, the development of multimodal attractor landscapes furnishes huge pay-offs for predictability. It is reasonable to assume that a system evolved to deal with change would be able to register such structures in appropriate cross-modal networks in the brain. That is, attractors in different sense modes will tend to become integrated into more inclusive, multimodal attractors, furnishing further powers of detailed predictability. In this way, information through one sense mode could readily disambiguate uncertain information through another.[34]

There is abundant evidence for such facility in almost all aspects of perception, cognition and behaviour. Humans, like other animals, are pretty good at predicting precise location from sound. It is clear, for example, that dynamic auditory stimuli, such as an approaching sound, can strongly influence vision.[35] Even newborn babies will turn their heads towards a sound. Although scientists have been aware for centuries that senses are used in concert, however, there has been little understanding of how this is achieved. But the fact that deeper spatiotemporal structures are involved is further suggested by several classic illusions. For example, the so-called McGurk effect occurs when a syllable presented audibly (spoken) is paired with a different syllable presented visually. This results in an audio–visual illusion consisting, not in the perception of one or the other syllable, but usually a completely different one. For example, a visual /ga/ combined with an audio /ba/ is often heard as /da/ – as if from a system striving to create a best prediction from the sample structure available in input.[36]

Another example is the way that activity or motion in one sense channel affects perception in another. In the familiar ventriloquist illusion, the visual stimulus of the dummy mouthing alters the perceived location of sound. Experiments have shown that if participants track a moving visual target with their eyes, a stationary sound simultaneously presented for about one second is perceived as moving in the same direction.[37] It is also a regular finding, as well as common awareness, that tactile sensitivity, as with touching an arm, is increased when the area can be seen. Indeed, a kind of 'super-sensitivity' to the touch occurs when the area is seen through a magnifying glass. Parts of the body not usually seen and felt together, such as the back of the neck, do not exhibit this effect.[38] All this, of course, is further demonstration of a new level of (cognitive) regulations emerged from the interactions between and within those at lower levels.

As might be expected, this demand for cross-modal integration, at a *cognitive* level, feeds back into activity at the neural network level. Multi-sensory cells, sensitive to more than one mode, have long been identified in the cerebral cortex.[39] In recording receptive fields of cortical neurons it is not uncommon to find some in *visual* cortex that are tuned to *acoustic* frequencies. It has been reported that cells in auditory cortex in humans are strongly activated by watching (silent) videos of speech-like facial movements.[40] It has also been noted how the 'strength of coupling' of signals from different modalities 'seems to depend on the natural statistical co-occurrence between signals.'[41]

Event concepts

Objects are rarely experienced in static poses. They are usually experienced as events in spatial transformation over time. Although each experience is novel there is a more invariant, general structure within. After internalisation, such structure, in the form of structural parameters, furnishes enhanced predictabilities about events. This is illustrated in studies of the phenomenon known as 'representational momentum'. Imagine viewing a moving object that suddenly vanishes and within moments you are asked to point to the spot where it vanished. When participants are asked to do this in experiments they always shift the vanishing point forward in the line of motion, as if predicting where the object would be those few moments after disappearance.[42]

Of course, this is simply one particular manifestation of the way you move and place your hands to catch a thrown ball: anticipatory movements that even infants have been observed to make in visually tracking moving objects. Dirk Kerzel showed that such predictability depends on some sort of knowledge of parameters of the event induced over several trials: that is, it does not occur with one-shot trials (as if a 'built-in' property of the system), but depends on what Kerzel describes as 'expectations building up across trials.'[43]

However, most objects are also experienced as events with other objects – birds with nests; bats with balls; people with tables and chairs and so on – in sequences of familiar spatiotemporal association. Such recurring events, with general form but specific variation, have been well studied by psychologists. Terms such as 'frame' or 'script' have been used to reflect a general pattern with varying contents.[44] For example, we may have constructed a script for getting dressed or eating at a restaurant. In each of these scripts there are said to be sequences or clusters of 'slots' being filled by predictable, though variable, objects or movements, depending on the particular context (e.g., getting dressed for work versus getting dressed for a jog). The general process has been called 'event structure perception.'[45] In neuroimaging studies these event boundaries appear to be reflected in changes in corresponding activity in sensorimotor brain regions.[46]

Action concepts

By the time more integrated structures like those above have emerged the system really consists of vast 'hyperstructures' arranged within an extensive 'hypernetwork', an increasingly complex attractor landscape. That idea is even more appropriate when we consider further integration with motor actions. Nearly all visual experience of objects is associated with motor action, even if it is just eye movements. Much acoustic input is also associated with readiness for action. Interestingly, this is shown in human musicians, in the way that listening to music activates motor areas of the brain as if ready for action 'based on higher-order rules of temporal organization'.[47] Not surprisingly, research shows how active exploration of objects enhances knowledge, and subsequent visual recognition, of them.[48]

Issuing a motor command for action on objects, say, isn't straightforward. It involves dynamic, moment-to-moment integration of inputs from somatosensory receptors in muscles and joints with other sensory data, as well as with knowledge background and the organism's current motivational state. Yet the body, like the outside world, is in constant motion and under constant perturbation. And the relations between body parts are also constantly changing with task requirements, injury, disease, development and ageing. Little wonder that Stephen Jackson says, the 'sensorimotor transformation problem' – maintaining exact motor action against a constantly moving sensory field – 'remains one of the least understood issues in motor control research.'[49]

It used to be thought that all motor-planning could be done by having a kind of 3D map or 'schema' of external space in the head, with the place of the body and limbs within it: a singular body-centred image. Recent research suggests that a far more *ad hoc* system appears to operate, in which 'movements appear to be planned and controlled within multiple coordinate systems, each one attached to a different body part.'[50] Each coordinate set seems to rearrange itself with the part as it is repositioned within its movement space.

The idea is supported by corresponding neural evidence. For example, there are neurons in parietal cortex sensitive to both sight and motion of limbs.[51] And there are others whose receptive fields are constantly being updated as if anticipating possible movements to objects within reach of an arm or a hand, or other part of the body. Some of these are the so-called mirror neurons. They respond with firing patterns that seem to reflect *intended* action, providing forward models of possible motor actions.[52] These and related experiments indicate a kind of continual and rapid remapping of space immediately around the body.

In other words, potential actions are continually being implicitly, and dynamically, constructed under the flow of constantly changing experience. This is what we would expect from a (cognitive) system constantly seeking to go beyond the information given in dealing with change, or even anticipating and making change. It is a mode of operation that is strikingly compatible with the dynamical view of a self-organising, multi-level, attractor landscape.

In passing, we may also note the deep interpenetration of mind, body and world implied in these discoveries. Kevin O'Regan and Alva Noë have reviewed various sources of evidence, suggesting that the integration of vision and action involves a more complex phenomenology than direct mapping from one to the other.[53] The blind man probing a manhole cover with the end of a stick has an impression of the object as projected 'out there', not in the hand and fingers where the sensations are. When riding a bicycle, you somehow become conscious of the whole texture and topology of the road surface (even when cycling in the dark), although all sensation is received by your hands at the handlebars, and through the seat of your pants. Through the basic cognitive functions described here, we can see how the outside world becomes 'as one' with the living mind and body.

Summary

A major aspect of the function of a cognitive system is, as Jerome Bruner, once explained, to go 'beyond the information given'. This function became an absolute prerequisite in an ever-changing environment filled with moving objects. A cognitive system is a whole, new level of evolved regulations, over and above, but emergent from, the regulations of neural activity. Neural actions, for example, deal only with two-dimensional patterns of light on the retina, whereas cognition induces, and deals with, the four-dimensional spatial and temporal structure emergent from it. Light patterns reflected from aspects of objects may consist only of moving arrays, whereas cognition induces from them complex features, whole object images, abstract concepts and the vast range of predictabilities that give them meaning, and so on. This is a far more complex intelligence-gathering process for an intelligent system than the epigenetic and developmental systems already discussed.

All of these studies suggest that experience is partitioned around deep covariation structures, rather than discrete features or rigid categories.

It is only that structure that makes them objects and renders them identifiable and predictable. More importantly, the research suggests that what develops in a cognitive system is really a vast system of attractors, based on structural parameters. These parameters specify interactions between variables and can predict 'most likely' variable values in variable circumstances. The attractors are themselves nested into overarching attractor landscapes, reflecting what I called 'hyperstructures' in experience. Perturbations from inside or out can tip those attractors into temporary chaotic states that drive a rapid search of state space for the best resolution. This is the signal passed around for motor action, further network attunement, or whatever.

Converting blooming, buzzing neural traffic into meaningful cognitive objects suggests how the whole dynamic, self-organising hypernetwork functions like a super-organ – a cognitive organ with its distinctive regulatory structure. And unlike previously evolved (biological) organs, this one can be constantly modified and updated. Little wonder that the emergence of the first cognitive systems created yet another evolutionary escalator, the complex properties of which we are still to properly grasp.

In this chapter, I have been talking only about the most basic cognitive functions and trying to explain how they have emerged from antecedent neural activity (itself emergent from prior developmental and epigenetic systems). Already, I hope, you can see that a crucial bridge has been created to the further vast complexity of cognitive systems. Let us turn to consider these in the next chapter.

8
Cognitive Functions

The previous chapter considered how animals deal with the main stimuli in their environments – that is, objects – to form concepts. In the process, a cognitive system – a whole, new level of evolved regulations – emerges from the regulations of neural activity. As a result, the cognitive system does not deal with the information delivered to it directly from the senses. It goes beyond the information given, to construct a cognitive world that, in turn, becomes an intelligent 'adviser' to further neural activities, and even to brain development.

All this becomes even more apparent when we consider the more complex cognitive functions emergent from the basic ones just considered, and their potency in the interpretation of reality and action upon it. There is abundance evidence, for example, of the way that learning, thoughts, feelings and motives affect perception.[1] And the importance of concepts has long been acknowledged. Jerry Fodor famously described the nature of conceptual representation 'as the pivotal theoretical issue in cognitive science; it's the one that all the others turn on'.[2] Stephan Harnad described concepts as the building bricks of cognition: that 'to cognise is to categorise'.[3] Beyond concepts, most cognitive scientists will think of 'higher' functions such as thinking, reasoning and problem-solving: the whole operation of an advanced intelligent system. But how, exactly, do these functions hinge on concepts? How can they be constructed from those building bricks? If we are attempting to find bridges between less and more complex living systems, this is one that we certainly need to focus on. That is what I will try to do in this chapter.

Fragments of cognition

Cognitive functions have been considered at a vast variety of levels. As described in Chapter 7, even the simplest nervous systems show cognitive functions. Integration of information, beyond elementary associations, goes on in the honeybee's brain with only 0.01% of the neurons of a human. It seems reasonable to assume that these functions expanded as animals evolved into more rapidly changing circumstances. But the nature of this expansion has not been well theorised as yet. And its reflection in, and the nature of involvement of, the brain itself, is still not clear. After all, many 'closed' behaviours, such as reflexes, instinctive routines, can be achieved with only rudimentary wiring and simple rules.

This uncertainty is not due to lack of research effort; but it could be due to choice of strategy, or underlying presuppositions, or both. In the face of a great imponderable, investigators have (quite reasonably) tackled specific aspects first. So the field of cognition has been broken down into specialised topics. By dint of funding and social policy, these have usually been ones of some immediate practical concern in areas such as education, child development, occupations or mental illness. These specialised topics today include learning, memory, knowledge, decision-making, inferential and deductive reasoning, thinking and logic. Each of these areas has thrown up its own questions, identified unique variables, cultivated unique models, and even explanatory philosophies. But little has been achieved in terms of construction of a general model of cognition, putting the role of the parts within a more global interactive system. As Martin Giurfa argues, 'Despite this diversity and increasing interest, a general definition for the term "cognition" remains elusive, probably because the approaches that characterize cognitive studies are diverse and still looking for a synthesis'.[4]

To help put things in perspective it is probably worth having a quick look at past and recent ideas concerning some of those topics.

Learning

Learning is usually considered to consist of changes in cognition through which past experience can inform present behaviour. For long periods in the history of cognitive theory, it was thought that learning consisted of acquiring simple associations from experience: what tends to occur with what in simple one-to-one fashion. Such associations, acquired by a kind of trail-and-error process, or even just from their repetition, could lead to expectations/predictions in future situations. Pavlov's dog hears the dinner bell and salivates at the expectation/prediction of food. 'Learning theory', based on such assumptions, dominated psychology for many years (and, in the human context, even came to dominate schooling).

A plausible mechanism for this kind of learning was proposed over a century ago; that is, changes in cell connections in brain networks due to effects of rewards in learning the association. This idea was taken up in the 1940s by Donald Hebb in his book *The Organization of Behavior*.[5] It suggests that stimuli or events associated in experience will become associated in the brain through changes in synapses, making the respective pathways more likely to be activated together in the future (it is sometimes referred to as 'Hebbian' learning). Writ large, across multitudes of connections, the pattern of firing and wiring is assumed to reflect the pattern of experienced associations. Hebb suggested that learning in this way leads to 'neural models' of the environment.

There is little question that such simple associations can be made in brain networks. The big question has been whether they can really account for more abstract (conceptual) learning, so demonstrable in all cognitive systems. It seems reasonable to ask, if this is all learning consists of, what has happened over the course of hundreds of years, especially in a rapidly expanding brain? Although we appreciate that the greater capacities of evolving brains permits better learning, Hebbian theory isn't clear about the ways in which human learning is different from that of apes, other mammals, or even bees, say.

This very conceptual simplicity also encouraged a rather limited research paradigm: decades of laboratory studies using repetitive, static stimuli, with animals and humans, in attempts to discover 'laws of learning'. But, as Noam Chomsky famously demonstrated, learning complex cognitive operations, as in human language, is quite different from learning simple associations. Output from a cognitive system, for example, is usually much more than mere repetition of experience: outputs (like inputs) are usually quite novel, in fact. In attempts to resolve such problems, sometimes two or more distinct learning systems have been proposed to coexist within the same brain. But it is still surprisingly difficult to explain how learning, even of simple motor skills, comes about.[6]

More recent neural network models have risen above the notion of simple (one-to-one, If-Then) associations. Instead they stress the overall configuration of connections, reflecting *patterns* of activity. These patterns are statistically deeper than simple associations, involve nonlinearities and so on. The models have shown how new levels of regulation can emerge: the historically developed connection configuration affects the activity pattern and the activity pattern, in turn, affects the connection configuration. This is a potentially creative, dynamic system for abstracting and utilising environmental structure, especially with the growing appreciation of external structural dynamics.[7] Of course, the specific network models we have seen so far lack the general capabilities of even the simplest real cognitive systems. Either way, the dynamical framework promises to transcend the traditional paradigm, and I will have more to say about that later.

Memory

Whereas learning has been viewed as the induction of new connection patterns, memory has been modelled as the storage of these so that they can be retrieved or used for comparison of past with present experiences in some way. The filing cabinet metaphor, and the storage of memories as discrete 'files' or 'traces', have been very prominent ideas. Models of memory have thus been preoccupied with proposing mechanisms through which the acquisition, storage, organisation and retrieval of associated elements – the 'memory trace' – are facilitated or inhibited. In the standard computational model, the use of memory is assumed to take place through a computational algorithm. This compares new inputs with stored patterns on the basis of their 'similarity' and then retrieves whatever further information is associated with it. With that, inputs are then said to become subject to further cognitive processing through the operation of further algorithms, under the supervision of a 'central executive'.

Again, the gap between theory and reality has created many doubts. The assumption of a static, time-frozen quality of experience is reflected in the use, in experiments, of fixed, simplified stimuli under stationary conditions. Word lists, digits and nonsense syllables have been very popular, as have static displays of pictures of objects, faces and events. These are presented to volunteers under various conditions, and responses evaluated, usually in terms of rate or accuracy of recall or recognition. As Richard Shiffrin, a leading memory researcher, explains in a review, the models are largely devised to fit and explain data based on those assumptions.[8] But as Karim Nader says, evidence from abundant sources now shows that memories 'are not snapshots of events that are passively read out but, rather, are constructive in nature and always changing...It is therefore strange that, although memory is dynamic and changing, the dominant memory model proposed to describe it emphasizes fixation. The two views are diametrically opposed'.[9]

Again, to meet theoretical conflicts, it has been suggested that cognition involves two or more different 'memory systems'. Shiffrin himself candidly admits, 'the work yet to be done far outweighs the progress thus far', and that current understanding 'is at such a primitive level that current models will in the not too distant future be viewed as comically oversimplified.'[10]

Knowledge

The nature of knowledge, how it exists and used in cognitive systems has been at the roots of psychology since Ancient Greece. Indeed the term 'cognition' stems from the Greek word 'gnosis', meaning 'to know'. The subject is now covered by vast literature.[11] Knowledge is the product or end result of learning, in which past experience somehow prepares us for rendering predictability in future experience. It is, of course, assumed to reside in patterns of neural connections: for example, in concepts and their interrelations. As David Elman and colleagues say, 'Knowledge ultimately refers to a specific pattern of synaptic connections in the brain.'[12]

But that doesn't indicate how knowledge enters into cognitive functions. Associationists have stressed how co-occurrences between attributes and/or events can be assimilated as concepts making up knowledge: for example, assimilation of the associations between 'fur', 'four legs', 'tail' and so on, can make up the concept of 'dog'. Such knowledge can, in turn, be used to make predictions from current to future experince. So-called *constructivist* theorists have stressed the internalisation of 'schemas', or internal models of experience. Such schemas go beyond simple copies from reality, such as feature-associations, to deeper relationships, but they have been criticised for their vagueness. Others propose that knowledge is a system of 'rules' that govern cognition. Nativist theorists, such as Noam Chomsky, have argued that the basic rules are 'innate', or genetically specified.[13] In his book, *How the Mind Works*, Steven Pinker mentions how psychologists feel 'perplexed' and 'baffled' about the nature of knowledge.[14] In his review of the subject, Emmanuel Pothos says that, 'Overall, there has not been a single dominant proposal for understanding general knowledge.'[15]

Thinking

It is generally agreed that at the core of cognition is a set of functions surrounding thinking, including problem-solving, decision-making, judgement, reasoning and so on. A vast variety of models of these processes has arisen from an impressive array of ingenious studies. In the first half of the twentieth century, the Gestalt psychologists demonstrated how the cognitive system appears to go 'beyond the information given' to impose order or organisation on raw sensory stimuli. Studies on animal and human problem-solving seemed to indicate constructive reorganisation of problem situations, such as to aim for good and complete form in the solution. The process seemed to indicate sudden 'insight', as when a chimpanzee would suddenly realise that inserting one stick into the end of another would make it long enough to reach food outside the cage. Of course, as suggestive as they are, such ideas remain rather vague.

The most common recent notion is that thinking consists of operations on internal cognitive 'models' or representations (usually based on concepts) in order to render some predictability from them. For example, Renate Barsch says that, 'Thinking is manipulating representations under evaluation, and thereby arriving at new representations.'[16] Michael Eysenck and Mark Keane suggest that thinking can be broadly described as 'a set of rules that manipulate representations to carry out thought process.'[17]

The work of Roger Shepard provided a suggestive glimpse of such functions. Participants were presented with pairs of line drawings of simple 3D shapes as if viewed from different angles and asked whether they were of the same or two different shapes. The time needed to make a decision was closely related to the amount of 'mental rotation' needed to transform one pose into the other.[18] The results offer a compelling impression of a nervous system creating an internal model that is then subjected to mental manipulations such as rotation.

The problem lies in specifying the nature of those representations and how they are set up and manipulated, by what kind of rules. These have been constant topics of debate. In the dominant computational model, thinking consists of processing of input elements (say a feature like fur) according to sequences of rules, as in a computer program. According to Steven Pinker, any problem can be broken down into a series of coherent steps. These can then be carried out by a computer program to form a machine 'that thinks', and so explain the vast repertoire of human thought and action.[19] So-called Artificial Intelligence grew out of the computational assumption that cognition is done according to fairly simple rules: we simply need to put those rules in a machine to get a machine that thinks.

As you might expect by now, the view has been increasingly criticised for the unlikelihood of many of its assumptions: for example, the tacit breakdown of continuous and coordinated experience into elementary features and well-defined steps.[20] Moreover, although internal representations are usually described by analogy with tangible images, such as pictures, maps, 3D models or even linguistic propositions, they cannot literally be like that. And it is doubtful whether creative processing can really occur through sequences of simple If-Then rules, which assume a repetitiveness that is rare in the real world.

All this may explain why the implementations of computational models in artificial intelligence have produced machines that do simple, routine things (such as selecting and counting items according to fixed criteria) very well, but cannot deal with fuzzy, dynamic inputs (such as moving images or human speech). As mentioned in the previous chapter, critics such as Jerry Fodor remind us that real sensory data in real life do not come in nicely carved chunks. Discrete characters for inputting to a keyboard, may be convenient for channelling into a computer program, but it's not what real experience is like.[21] So those programs are unlikely to yield truly realistic models of thinking.

Laboratory studies of thinking, on animals or people, have also tended to focus on very simple, well-defined, stationary (non-dynamic) tasks devoid of context. Studies of 'logical' thinking, in particular, have used very simple syllogisms in which determinate inferences have to be drawn from given premises, for example,

All A's are B's
Are all B's A's?

This strategy has probably been self-defeating because, as Michael Eysenck says, while psychologists have focused mainly on artificial, well-defined problems, most everyday problems are ill-defined.[22] Consequently, we seemed to have learned a lot about the difficulties humans face with unreal cognitive tasks in an unreal world, but not so much about real thinking in the real world.

One unfortunate consequence of this strategy seems to have been a remarkable underestimation of human thinking/reasoning powers in the real world. This pessimism about cognitive ability of both adults and children is, on the one hand, strongly reflected in so-called reasoning tests (or IQ tests). Studies of problem-solving in real-life contexts, on the other hand, reveal complexities of thinking that greatly exceed those manifested in such tests.[23] As Eysenck says, 'There is an apparent contradiction between our ability to deal effectively with our everyday environment, and our failure to perform well on many laboratory reasoning tasks.'[24]

We have probably also underestimated the 'thinking' abilities of non-human animals. Research over the last few years has revealed previously unsuspected complexity of cognitive processing in invertebrates such as bees and flies, as well as in vertebrates such as fish and jackdays. Much of this was mentioned in Chapter 7. It seems clear that current models of thinking are failing to capture the true qualities of cognitive operations in complex, dynamic situations, from the foraging of the honeybee to humans carrying out cooperative, constructive acts over long time periods.

Problem-solving

Most of animal cognition is problem-solving, usually defined as working out how to move from a current to a desired goal situation. Again for strategic reasons, most research has used highly simplified, unnatural, laboratory tasks or puzzles in a strategy that has probably been counter-productive. In humans, game-like scenarios such as the Tower of Hanoi (involving blocks to be arranged in a certain order on rods) were presented to participants who then answered protocols on cognitive routes to the final goal. In animals, tasks such as finding ways around mazes, or out of cages, and obtaining food have been used.

Only over the last two decades have researchers realised that problem-solving processes differ with contexts, across knowledge domains, development and levels of expertise. The implication is that findings obtained in the laboratory, or from context-free tests, don't necessarily generalise to problem-solving situations outside. The aim of a global theory of problem-solving has, in fact, been relinquished as researchers have turned to investigate problem-solving in different 'real world' domains (e.g., in school subjects or work situations) and the factors that influence it.[25] The result is that there is still no coherent model of problem-solving as a cognitive process in either animals or humans.

The mystery of cognitive functions

I hope this brief treatment of particular topics is enough to indicate how theorising about 'higher' cognitive functions remains disappointing, in spite of considerable effort and ingenuity being applied. In my view this is probably because the function of cognition is itself misunderstood, largely, in turn, because the nature of experience is misunderstood. We get rather mechanical models of minds doing relatively simple things in rather fixed, static environments: so much of the research and theorising is based on false presuppositions. As mentioned in Chapter 1, cognitive psychologists tend not to have appreciated the depth of structure of natural experience, or to have claimed that it is rather rare. In what follows, rather than fixating on specific aspects of cognition I want to show how recent theorising from a dynamic systems perspective is helping provide a radically new perspective on cognitive functions in general.

From deep structure to cognitive functions

What the brain and its cognitive system need to be intelligent about is a multivariate, constantly changing stream of experience. By virtue of its deep, spatiotemporal, structure this experience is captured in brain networks as an evolving attractor landscape. Each attractor condenses clusters of statistical structure among connection weights – parameters that are used to perform optimisation functions, such as creating best fit to current inputs and shaping optimum responses. The attractors, generally, are maintained on the edge of criticality by constant perturbations from inside and outside the brain. We have seen that, under influence from top-down activations, a sudden switch to chaotic states helps find optimal resolutions to current inputs very rapidly.

As the term 'landscape' implies, though, attractors do not 'act' in isolation, they interact with interesting consequences. It has long been known that large ensembles of such attractors, activated together, undergo transitions as more inclusive clusters of networks, or 'networks of networks', operating together, with important emergent properties.[26] We saw in the previous chapter how such clusters self-organise in various ways: integrating primary structures to form images of whole objects; achieving multi-sensory integration; forming event structures and so on. It is now evident that separate attractors can sporadically integrate into a whole global attractor system to form the basis of higher cognitive functions.

This tendency of networks to collaborate, to yield further emergent properties, has been extremely important in the evolution of complex cognition. It parallels, at a higher level, that in the evolution of epigenetic regulations. In the latter, the genetic functions of protein synthesis have been assumed into, and modulated by, dynamic epigenetic regulations, and these have been assimilated into a dynamic physiological system. Here, the cognitive products of the activities of initially distinct neural networks become integrated to regulate what goes on within them. Such cooperative/collective dynamics between previously separate attractors are now being used to explain complex forms of intelligence and behaviour, from insect societies to brain activity in cognition.[27] In each case, the result is a coalition of cooperating attractors self-organising to optimise functions. Because they are important to the understanding of higher cognition; it is important to be clear about the extent of these brain networks, and how cognitive functions reflect whole brain processes, with intense and rapid intercommunication between centers. Let us look now at some of that interaction.

Coalitions of networks

Outwardly, nearly all cognition and action in animals involves a seamless melding of sensory and motor sequences. Hunting, for example, consists of scanning the territory, spotting, categorising and identifying objects, tracking the prey, assessing the feasibility of catch, approaching, catching, killing, preparing and eating. Inwardly there is another, more complex, flow of events: reception, assimilation and categorising of sensory inputs; internal 'feelings' arising from bodily states; motivational states (such as hunger or exhaustion); activation of knowledge and memory states; registering of somesthetic signals from joints, tendons and muscles, creating images of current bodily dispositions; creating the spatiotemporal pattern of activities to muscles leading to action (which, in turn, changes the stream of sensory and other experience). All this, of course, involves intense communication between brain centres.

As mentioned in Chapter 7, signals from the distance sense organs (sight, smell, hearing) converge and interact with those from taste and touch. These inform the animal about current context with the help of feedback from centres in the cerebral cortex. But all sensory outputs/percepts, including those from muscles and joints, are sent in parallel to subcortical centres concerned with the other important context, the current internal state of the animal itself. Indeed, even the smallest functional brains, as in flies and bees, have centres monitoring the internal milieu, and activating neurosecretary cells to modulate it.

Prominent among these centres in vertebrates is the hypothalamus that receives nerve impulses from all sensory systems. It also has rich beds of chemoreceptors monitoring the internal state – the homeodynamics – of the body. It responds to environmental perturbations (as signaled from the senses) by issuing certain signals to endocrine glands (as well as creating many other aspects of feelings). Much of this is done through the neighbouring pineal gland which, in turn, secretes a wide range of hormones into the circulation to influence functions elsewhere in the body, like the release of adrenaline in readiness for muscle action.

In addition sensory channels share rich reciprocal connections with a number of other subcortical networks in the limbic system (e.g., the amygdala), also concerned with monitoring and modulating feelings. There is also an intense interplay of the hippocampus and amygdala in interrelating current sensations, emotional states and motivations, all influencing a 'goal-orientedness'.[28] This active stance becomes manifested in the process of 'preafference'. The process is created by discharges from the limbic system (including the entorhinal cortex and the hippocampus in mammals), to all sensory centres in the cortex, in order to maintain a continuous expectancy of changes to come, and attention to them.

The outputs of these limbic-emotional areas, that is, provide the brain with an *affective* image, or 'feeling', that characterises the active motivational state. This promotes, inhibits or helps shape responses in brain networks, affecting numerous aspects of behaviour.[29] Stimuli may be quickly identified as potentially significant or otherwise by these affective mechanisms. These, in turn, amplify the processing, bring additional attention mechanisms to bear, and so help to shape the percept.[30]

The activity, so far, may result in learning, or adjustment to connection configurations, but the actual shaping of a motor response is just as likely to be part of the flow. The motor output is a spatiotemporal pattern of activation from brain centres in the cortex, through the basal ganglia. These are sets of neurons at the base of the forebrain, strongly interconnected with other centres in the thalamus, and thought to be involved in response selection and timing. Responses are then sent (in their spatiotemporal patterns) to motor fibers to the muscles.

But another aspect of the preafferent 'copy', now of the motor pattern, is also sent, simultaneously, to all of the central sensory systems. The copy prepares them for the impending changes in sensory input that will result from the motor action. Doing so ensures that even simple decision-making is being planned in anticipation of an action. This, too, involves further extensive interactions between centres. As mentioned in the previous chapter, networks in cortex are constantly being prepared to anticipate possible movements to objects or other stimuli within reach of a part of the body. At any one time, the position of a limb, say, is being monitored in one part of the brain (the parietal cortex) by integrating proprioceptive signals from another (the sensorimotor cortex) with the 'efferent copy' signals from motor cortex.[31] The result is a prediction-action cycle adaptable in rapidly changing situations demanding rapid response. In escaping from a predator, for example, these interactions produce a reasonably accurate prediction of the disposition of the attacker a few milliseconds in advance, allowing a forward movement plan of avoidance.

So far, it may seem that what we have been discussing is the flow of neural signals. It is important to note, however, that this complex traffic flow is carrying *cognitive* constructions – already the results of attractor activity – over and above the neural activity. This is clear in a number of observations. For example, responses of the amygdala are more closely related to the percept – what a stimulus 'means' to the individual – than to the physical characteristics of the stimulus itself.[32] Likewise, in numerous areas of brain, neuronal activity has been found to vary according to the size of the reward being expected of an activity – that is, from previous experience – and/or to the current degree of motivation. Reciprocally, anticipation of a more valued reward is reflected in measures of arousal, attention and intensity of motor output.[33]

Accordingly, what is being processed, at a variety of levels, are, already, cognitive interpretations, not passive patterns of neural activity. This echoes the findings of Freeman and colleagues that it is not external smells *per se* that animals respond to, at least directly. Rather they respond to internal activity patterns created by the chaotic dynamics within the olfactory bulb.

It is also worth noting that, although feeling/emotion used to be considered to be 'the enemy' of cognition, the degree of collaboration between sensory, affective and other cortical centres suggests otherwise. All cognitive activities have an emotional aspect, and emotional value. This is, in fact, what 'embodies' and contextualises the cognitive system, and readily distinguishes it from the 'cold' computations of a mechanical robot. It is also vital for learning and creativity. Knowledge, for example, emerges as 'more general concepts at higher levels. In these connections diverse knowledge acquires emotional value. This emotional value of knowledge is necessary for creating more knowledge'.[34]

This is just a mere glimpse of the richness of interactions in brain/cognitive networks. It is now clear that sensation, feelings and motor-planning do not act in isolation, with cognitive processing passing from one to the other, as if in a factory. Rather it emerges, self-organising, from an integral, 'network of networks'. These provide rich foundations for the developing attractor landscape to create cognitive functions. The speed and creativity of such dynamics stand in stark contrast to the more limited abilities of traditional models of cognition. For example, motor-planning absolutely demands non-linear dynamics because the form of a motor action can rarely be specified in advance. It needs to emerge from reciprocal dynamics among different networks. In the traditional models the network was 'told' the steps to follow for making a decision and forming the action, in linear fashion.[35]

But there is another crucial aspect of this arrangement. Cognitive systems became critically demanded in environments in which the informational structure is continually changing throughout life. Only the kind of network collaboration we have been sketching here can furnish the dynamics needed to constantly assimilate new structures and respond fruitfully to them. Luiz Pessoa says that brain networks, and their activities, are so integrated as to form 'dynamic-coalitions' of brain areas that rapidly coordinate in a 'coupling topology'.[36] I now want to discuss how such dynamic coalitions foster the continual development of cognitive constructions into more abstract, and more complex, forms.

Piaget and reflective abstraction

The whole evolutionary distinction of cognitive systems, compared with previously evolved systems, is that they continue to change themselves throughout life, assimilating ever-wider and -deeper structures and predictabilities about their worlds. Perhaps a good way of looking at these powers of cognitive systems is through the eyes of developmental psychologist Jean Piaget. In the early decades of the twentieth century, Piaget was an epistemologist, interested in the nature of human knowledge, and how it changes over time (as in, particularly, the history of scientific knowledge). He viewed 'logical' thinking, as exemplified in scientific method, as the summit of both evolution and cognitive functions, and argued that it could best be understood by studying the development of knowledge in children. Piaget viewed knowledge development as an advanced form of the epigenetic adaptations found in animal life generally. But, unlike the maturation of pre-adapted physical processes of the body, adaptation at the mental level cannot be predetermined. The evidence for this, he said, is that, in the emergence of knowledge and reasoning powers in individuals, 'there is a continual construction of novelty'.[37] It is this (ultimate) creativity or inventiveness of cognition that stands out as the strongest aspect of Piaget's theory.

Piaget saw knowledge (and with it cognitive function, or intelligence) as consisting of operatory 'schemas' induced from experience through action on the world. These schemas, he said, do not reflect independent dimensions of experience, but the dynamic coordinations among them. For example, a child rolling out a ball of clay is experiencing various coordinations: changes over time between the length and thickness, between the visual appearance and the motion of the ball, and between these and the sense receptors in muscles and joints. In successful development a whole complex of coordinations is being 'internalised'. When completed, the schema renders future action of the same type more predictable and skilful. The world of the infant may be a 'blooming, buzzing confusion', but action soon begins to expose those coordinations and render the world more orderly and predictable.

When the coordinations have been internalised, representations in mind are in *equilibrium*, or perfectly corresponding with the coordinations in experience. It is such equilibria that Piaget saw as the developmental prerequisite for logical and scientific thinking. All the time, though, the equilibrium is being disturbed by new challenges making consequent re-equilibration an ongoing process, leading to schemas with ever-deeper representation of reality, and ever-wider applicability – the essence of development. For example, from the initially undifferentiated sensory stream, the infant soon begins to distinguish objects as entities distinct from one another, and from self, and to form schemas of their relationships or inter-coordinations. These permit an early level of logic in the interpretation of events in the world and of action upon it. One example is the understanding of means–ends relationships, whereby one object can be used to move another.

Note that the schemas are not, themselves, passive reflections of reality, however well they assimilate and represent its coordinations. They interact with each other, such as to form more integrated schemas. For example, the concept of 'object permanence' – permitting the prediction that objects persist even when they cannot be directly seen – is a higher schema reflecting relations *between* objects, not just properties of objects themselves. A small perturbation at the local level, such as the disappearance of one object behind another, has led to reorganisation (accommodation) of an existing object schema in relation to others, with emergence of a more encompassing one at a higher level.

This means that the system now acts on a new, conceptual level. Cognitive functions, and new knowledge, can be exercised on a level detached from actual sensory experience in the here and now. This became evident in the vast variety of famous 'tests' Piaget used with children of different ages. For exampe, in late infancy, one object can be used to 'stand for' another, as in symbolic play. Later, reality can be more easily divorced from superficial appearance, as when a child will realise that the two rows of counters below can contain the same number (Figure 8.1). Whereas the first schemas are compressed 'copies'

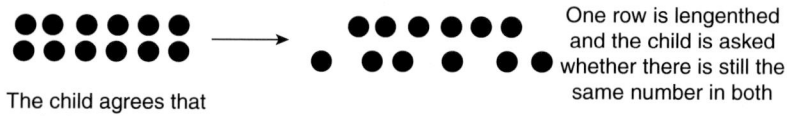

Figure 8.1 Conservation of number over changes in appearance

of reality, the more developed schema allows representations beyond immediate appearance, and thus to 'anticipate' novel possibilities, as in maths and theories of science.

This is Piaget's important concept of 'reflective abstraction': the idea that knowledge can arise from interaction among the schemas – perhaps because of some initial inconsistency between them – to obtain a more abstract level, detached from immediate experience. One example is the formation of higher inclusion classes (e.g., fruit) from simple concepts (e.g., apples, strawberries). The latter are 'empirical' abstractions, in that they form from what is directly experienced. But the inclusive class demands a higher self-organisation, over and above what is experienced. And it takes time to develop. Younger children can answer an empirical question such as 'are there more apples or strawberries?' because they can be counted. But he/she may fail the question, 'are there more apples or fruits?', if the classification schema, and the logic that goes with it, has not developed. This demands more than 'empirical' abstraction, derived from direct experience (as in the integration of visual features into concepts, described in Chapter 7).

Another example is that of 'transitive inference'. The conception that object A, say, is bigger than object B, may be gained from direct empirical experience. Likewise with the experience of B being bigger than object C. But appreciating that Object A is necessarily bigger than Object C requires reflective abstraction, and the emergence of a new schema that provides an important system of logical thought. Both examples demand abstraction of the more implicit structure, beyond what is available in direct experience or appearance. This creates a higher conceptual level with new 'logical' powers crucial, in humans, to maths and science.[38]

The emergence of reflective abstraction provides a powerful means by which complex intelligence can generate much *more* complex – and more powerful – intelligence from its own activity. It is, of course, a manifestation, in real cognitive systems, of the effects of the synchrony among attractors – or attractor coalitions – described in the previous section. This self-organising ability of cognitive systems, yielding knowledge about the world deeper than that in immediate experience, must have been a tremendous boost to the evolution of predictability and adaptability.

It would be interesting to know when, in the course of evolution, it did emerge. Many studies on animal cognition have tried to assess the appearance of such cognitive powers in non-human species. Although requiring ingenious modifications of Piagetian tests, the results for mammals and primates have not been clear cut.[39] It is clear that a variety of species have some concept of object permanence. For example, dogs will look for hidden objects; and squirrels and even birds such as jays will store and retrieve food from caches. Also, the logical ability of transitive inference has been reported in a variety of birds and mammals (an example is given in Chapter 9). The fact that bees and other insects can form general concepts also implies the presence of a kind of reflective abstraction even in the earliest brains. In his book *How Animals Think*, however, Clive Wynne duly warns of the dangers of anthropomorphism and of the importance of assessing the cognitive abilities of particular species within their ecological contexts.[40]

Reflective abstraction in networks: other evidence

In fact, within non-linear dynamic systems, some kind of reflective abstraction, and emergence of higher functions, is not a surprise. Many investigators have observed how more complex, adaptable, structures readily emerge from interactions among initial attractors, through evolving hierarchical levels.[41] We saw in Chapters 4 and 5 how more 'abstract' gene regulation modules emerge among signaling networks at several hierarchical levels. The whole of morphological development, including cell fate, emerges as modules or high level attractors among lower level interactions. As we shall see further in the next chapter, interactions among 'dumb' individual ants or bees lead to complex social networks, obeying new properties and laws at levels higher than any individual. It is hardly surprising that evolution of higher cognitive functions should build upon such principles.

The brain, however, appears to be a dynamical system of unprecedented complexity. Neural activity interacts within and between numerous levels to permit huge scope for emergence of higher functions by reflective abstraction. Initial attractor landscapes are set up by direct ('empirical') experience with different classes of stimuli, as described in Chapter 7. But these interact as coalitions of attractors, on several hierarchical levels, establishing increasingly abstract (cognitive) principles of regulation.

A lot of suggestive evidence for the process comes from the use of ANNs. Even quite simple networks give rise to higher-level regulations among interacting nodes and their attractors. Studies with ANNs show how continuous perturbation leads to drastic changes in the network dynamics and structure, and 'can raise new dynamic regimes', especially where there are recurrent connections.[42] Of particular interest are experiments in which ANNs are 'coupled' (i.e., with reciprocal connections), much as they are in real brains. Kunihiko Kaneko showed that, as the numbers of units in such networks increases, clustered and hierarchically organised activity is generated and evolves over time.[43]

In dynamical terms, the shape of input attractors are refined, or error-reduced, by feedback from 'higher' attractors, which themselves become reshaped by new feed-forward processes. These emergent states efficiently constrain and guide the flow of activity through the attractor landscape, achieving high level 'cognitive' functions. These higher attractors afford excellent perception and categorical prediction from sensory inputs.[44]

Finally, there is other psychological evidence for reflective abstraction. In the previous chapter, I mentioned studies on childrens' strategies in solving simple maths problems. It was shown how, with experience on the problem, new regulations developed through self-organisation of the initial attractor structure. A similar sort of emergence seems to occur in language learning. For example, learning the word sounds – the phonology – of a foreign language does not take place as a simple 'add on' to the native language, as a new, independent attractor system. Rather, it also seems to modify perception of the native one, even changing speech in the process. 'Thus the pre-existing phonological organization is malleable. Learning does not entail simply an addition of a new category but in fact changes the existing attractor layout'.[45]

This parallels studies in motor development. Motor control seems to progressively emerge as higher levels of coordination from initially lower level routines. Esther Thelen and colleagues followed changes in coordination of supine kicking in human infants from 2 weeks to 10 months. Patterns of extension and flexion in hip, knee and ankle seem to be originally coupled, limiting variability of movement. This coupling decreases drastically after 8 months, however, permitting individual joint actions and thus new coordinations. Similar development is seen in inter-limb coordination, and again, new forms of higher coordination become possible.[46]

All of this emergence of increasingly complex functions suggests fundamental principles shared by evolution, development and cognitive systems. The latter are distinctive, though, in that the higher functions emerge from attractor coalitions with remarkable speed, complexity and adaptability. This must have put cognitive systems on another evolutionary escalator. Perhaps the best evidence for that is the accelerating expansion of brain size across evolutionary time, as if suddenly finding great selective advantage in network capacity: the bigger the network capacity, the more abstract the regulations that could form.

New view of cognitive functions

Traditionally, cognitive functions have been modelled as if a fixed set of mechanical processes. Failure to adequately conceive the relationships between cognitive systems, evolution and the nature of experience, has meant that they have been studied in piecemeal fashion, each 'piece' with its own rules and principles. But, the new view stresses both the unity and adaptability of cognitive processes, reflecting those in the brain from which they emerge. Because of the demands placed upon them, we have seen how cognitive systems do not arise from genes or from experience (or from a 'blank slate'): they come to 'make themselves' through development, throughout life, limited only by the evolved network capacity of the brain. The increasing complexity of dynamics in evolving networks have allowed organisms to predict, anticipate and 'make' the future with ever-increasing depth of cognition and action. This seems to be the main theme in the evolution of cognition.

It follows that we will never understand cognitive functions without relating *particular* aspects of cognition to the whole. Instead of special processes within toy topics, with assumed built-in rules, we need to understand how the coalitions of interacting attractor states develop their own operating rules in ever-changing contexts. Of course, some structural and functional parameters, pertaining to more stable aspects of the environment, will be set up by developmentally canalised processes (see Chapter 5). For example, the sensory systems are 'pre-tuned' by canalised development to the stable wavelengths of light or of sound that we need to be sensitive to. But that depends on the environment to be adapted to. Likewise, the reliable physics of sensory transduction and nerve conduction mean that the basic arrangement of neurons and connections in the cerebral cortex can be set up in a consistent pattern, also by canalised development. With regard to more changeable aspects of environments, however, other connectivities must emerge in self-organised fashion from actual experience.

Generally, there is a close relationship between the relative plasticities of brain structures, on the one hand, and the kind of experiences an animal has to cope with, on the other. And that relationship will be reflected in the size of the brain area. The part of the insect brain called the mushroom body, thought to be involved in learning and memory, has a basic cellular structure that develops in early life, but also shows a great deal of structural plasticity, as might be expected for foraging animals. The structure is much larger in the honeybee than in the fruitfly, however, supporting the former's interaction with other brains as a social insect, and the need to form higher levels of regulation. The trade off between stability and plasticity depending on ecology is seen throughout evolution.[47] But the overwhelming trend in vertebrate evolution is the vast expansion, and connective plasticity, of the forebrain concerned with the integration of inputs and creativity of behaviour.

As researchers have explored the nature of that plasticity, it has been increasingly realised that organisms' relationships with their world are more intimate than used to be realised. This has led to several recent ideas about 'extended' cognitive systems. From the first living systems, living processes have internalised the external environmental structures through a number of devices. But the external structures are usually ones already affected by the activities of the organism in a continual reciprocal relationship. Accordingly, cognitive activities arise over neural states, over bodily states, *and* over the structure of objects and events in the surrounding environment.[48] Another way of putting this is to say that our minds are 'bigger than our brains'.[49] David Pincus, in discussing the fractal (i.e., deeply structured) nature of brain functions and cognition puts this, perhaps hyperbolically, when he says that 'everything in life is connected and that all of the universe is alive within these connections.'[50]

What I have stressed, here, is the importance of one particular aspect of animals' relationships with their environments in evolution. This is the emergence of reflective abstraction as the basis of 'higher' cognitive systems, and of acquiring deeper representations of the structure of the world. It constitutes another crucial bridge in the understanding of the complexity of mind. The idea is also important in putting cognitive science on a firmer footing. As Lynn Andrea Stein says, 'We are standing at the cusp of what Thomas S. Kuhn called a paradigm shift, in which the very foundations of our field are being reconceived', and that, '[i]ncreasingly, a more communal, contextual, interactive approach to cognitive science is coming into its own.'[51] In the rest of this chapter let me just consider a few implications of this view.

Dynamic learning

One of the symptoms of inappropriate assumptions in cognitive science has been the way that relationships between learning, evolution and development have never been understood. They have almost, at times, been viewed as antagonistic aspects of cognitive systems. Now they fall into place. Evolution has selected developmental plasticity as an adaptable strategy for dealing with changeable environments. Learning is the same thing as *lifelong* developmental plasticity (in the brain) for dealing with lifelong environmenatal change.

Appreciating why learning evolved provides new insights into its nature. Yes, it consists of modification of connections in nerve networks as theorised for most of the last century. But the modifications must be such as to abstract the non-linear statistical structure in current experiences. Accordingly, what is acquired are not independent, additive adjustments to synapses, but ones that capture the structural parameters of experience. The result renders the current and future situations more predictable. In dynamic terms, we say that processing of inputs creates or updates attractors in extensive attractor landscapes. These interact as attractor coalitions to provide the basis of reflective abstraction – and also more complex cognitive development

Some of the learning/development results in specialisation among the attractors, the most obvious ones relating to different sensory inputs: 'learning establishes attractor landscapes in the sensory cortices, with a basin of attraction for each class of stimuli that the animals have learned to identify. The basins of attraction are continually re-shaped by experience, and each attractor is accessed by the arrival of a stimulus of its learned class.'[52] But there are also more central regions of brain in which the intrinsic wiring is more suitable for handling some informational structures better than others. For example, in the mammalian brain, the initial configurations of the hippocampus seem to be specially adapted to integrate 'spatial' information, or where things are in relation to one another. Areas in the inferotemporal cortex seem to have evolved architectures that support the processing of finely timed acoustic information (and, later, human speech).[53] 'Higher' learning is the more general abstraction from the separate contributions of these specialised areas.

Note that what is learned – the updated configuration – is much more than a mirror reflection of the outside world. It captures the non-obvious depth and structure behind it that furnishes predictability. And it includes 'action' and 'affective' components, and, of course, emergent aspects that have not been directly experienced.

Dynamic memory

As Sylvain Chartier and colleagues point out, nearly all models of memory treat information storage and retrieval as of fixed point attractors (fixed files or traces), ignoring other possible dynamics.[54] In fact, there has not yet been a convincing NLD model of memory, and much yet needs to be explored and revealed of this most intriguing of cognitive functions. Certainly, we can say that memory is the series of updates in network configurations, and not the multitudes of discrete, traceable 'engrams', proposed in traditional associationist psychology. As described earlier, the configurations (of synaptic 'weights') must capture spatiotemporal covariation parameters, rather than specific values of independent variables. The flow of experience is thus 'recorded' as sets of incremental adjustments to these parameter values across the arrays of synapses. In the attractor landscape those parameters serve as the 'grammar' from which specific expressions can be recalled, or against which familiar experiences can be recognised. Activation of one parameter value will tend to activate all others in the original configuration, so reinstating the original memory. The fact that virtually any memory will be embedded in complex event structures, also involving other objects, sharing a veritable swarm of structural parameters, will also help memory activation (just as a photograph of a familiar place or activity, say, can revive a vast diversity of memories associated with it). Since the parameters are spatiotemporal, so the reinstatement is a 'movie-playback' of a spatiotemporal sequence. All this is consistent with ANN learning in which 'memories' are clearly overlain (superpositioned) within the same set of unit synapses with 'weights' that are clearly nonadditive (i.e., multiplicative).

Familiar objects or events are received and interpreted as familiar combinations of parameter values. These can evoke just those parameters in the memory attractor, which, through temporary chaotic mode, can re-instate the original memory. The model of Chartier and colleagues indicates just such dynamics in recovery of memory from fuzzy inputs. As a very simple illustration consider the well-known logical function known as the EXCLUSIVE OR (or XOR) problem. In this problem a network (or a real person) is presented with pairs of numbers that fall into

Table 8.1 The XOR Problem

Number Pairs	Category Output
0 0	0
1 1	0
0 1	1
1 0	1

one or other of two categories, as shown above (Table 8.1). The task is to learn the categorisation correctly, until, given any one of the pairs, the network will always produce the correct category (0 or 1). It is quite easy to train a simple network to do it by repetitively presenting it with the 'inputs' and keying in the feedback, which adjusts the connection weights until performance is satisfactory.

As you may be able to see, the solution cannot be learned by a simple, 'linear', association or correlation (e.g., between the category value and the value of any one of the pairs). It requires abstracting the deeper structural parameter through which the association between any two variables is conditioned by the values of the other. This is easily assessed statistically, but must be memorised in the network in terms of the sequence of adjustments to connection weights. In the process it is storing the parameter values and the deeper structure. Subsequent presentation of any of the possible pairs then reactivates that step in the sequence, in turn generating the permitted configuration containing the correct category value.

Real experience, of course, may consist of dozens or hundreds of variables, with many structural parameters, usually involving time. This is why recognising an input subsequently – or regenerating the best fit to inputs – from a myriad possible combinations may require temporary chaotic activity within the network.

Memory as structure captured in a configuration, rather than discrete files or traces stored as independent variable values, explains several prominent findings in memory research. With additional learning, and parameter updates, memory attractors can become modified and reorganised in relation to one another, such as their shared structural parameters. This explains why semantically related memories tend to be recovered together. For example, a person being asked to recall items in a category such as 'animals', will tend to recall in groups of related species: all the birds, first, say; then all the fish and so on. It explains why events will tend to be assimilated into an existing parameter structure, and then be recalled in a form closer to a personal or cultural ideal.

This harks back to work in the 1930s by Frederick Bartlett. He described studies in which participants were asked to reproduce stories and pictures experienced earlier. They typically elaborated or distorted the originals in ways that imposed their personal and social conventions over the original storyline. Bartlett thus concluded that remembering is a reconstructive, rather than merely reproductive, process, involving memory 'schemas'. In descriptions that would not go amiss in a dynamic framework he said: 'It is with thinking as it is with recall: Memory and all the life of images and words which goes with it are one with...the development of constructive imagination and constructive thought wherein at length we find the most complete release from the narrowness of present time and space'.[55]

Learning and memorising, in other words, are different aspects of a unified process of adaptability, pervading the system as a whole. As Velichkovsky says, 'there is no single brain structure which could be claimed to be one of the memory modules'; neural activities, and the cognitive functions arising from them, 'are multitasking geniuses rather than narrow-minded "idiots"'.[56]

Dynamic knowledge

Knowledge is the sum total of network attunements residing in the attractor landscape at any particular time. Animals or humans only truly know a domain of experience when they have assimilated the deep structures that make events within it more predictable. True enough, it is *possible* to describe even the simple associations that have figured in many models as a kind of knowledge. But these are shallow reflections of real experience, and of the depth of knowledge that is needed in cognitive systems, even in the primitive ones in invertebrates. Experience of even the simplest visual feature – that of a line – is, to the visual system, structural in form, and is, therefore, material for a kind of knowledge: knowledge that becomes vastly more complex and providential as it becomes increasingly embedded in deeper structures.

An important aspect of reflective abstraction is that interactions within attractors, under constant perturbation, frequently reorganise to form new emergent structures of knowledge. These include hierarchies of superordinate concepts, multimodal concepts, novel abstractions, mental or logical structures, sudden new insights and so on. As we will see in Chapter 10, this emergent potential became much expanded in humans, transforming the nature of knowledge in the process.

Dynamic thinking

Thinking – the 'core' of cognition – in animals generally, consists of the trajectories of activations through the attractor landscape. The trajectory may be impelled by a number of factors: hunger may initiate a planned search for food; dissonance or contradictions between current attractors may need to be resolved; or a means of escape from danger rapidly devised. In each case, the pattern of neural/cognitive activity is drawn into compatible basins of attraction, some of which will be higher-level attractors created through reflective abstraction. Activity within and between attractors may then create a novel conception for action. This may, in turn, emerge as motor patterns; further adjustments to specific attractor basins as a knowledge update and so on.

Thinking is inevitably embedded in the individual's ecological context, operating as an 'extended mind', as just mentioned. Interactions between that context and the internal attractor system may lead to thinking that actually delays decisions, allowing responses to evolve over longer time periods. This is obviously useful in stalking and hunting in animals, or in many important decisions in humans. Thinking activity may also continue progressively, perhaps as the result of gradual re-attunements, as when we slowly change our minds; or suddenly, as when we gain insight into a situation, understand the meaning of a novel signal from context, or reach a categorisation through our concepts. Or it can oscillate between rival states, as when we can't 'make up our minds'. All these aspects of thinking present traditional models of cognition with difficulties, but are readily explained within a dynamic framework.

A crucial aspect of thinking within dynamical attractor landscapes is its rich creativity. This is not really explained, and rarely touched upon, in current cognitive theory. However, creativity is a natural property of non-linear dynamics.[57] By definition, changeable environments demand creativity of responses from living systems.

Again, though, we need to remember the role of affective factors in cognition, and that each state of a cognitive system is tied up with a feeling. When a familiar object or face is recognised, say, positive or negative feelings are part of that recognition. This meshing of all cognitive activity in feelings – and, through that, their embeddedness in the outside world – is what distinguishes animals from robots. It resolves as the deeper phenomenon of belief, which might be defined as knowledge with feeling, together promoting confidence (though with variation in balance between the two). We can talk about robots learning, memorising, making decisions, having knowledge and so on. But we don't talk about robots having beliefs. Without it they remain less creative than the humans who invent them.

This further clarifies the meaning of 'meaning'. All that can be predicted about a perceived object, event, thought and so on, in association with feelings, is the definition of meaning. This, too, is an immensely creative property that virtually defines the individual. The attractor landscape made up of the cognition-feeling complex *is* the individual identity. In humans, it forms the concept of self, known by self and others, defining all sorts of response tendencies and so on. The origin of this identity of a cognitive system in its self-organising, dynamic, properties, finally banishes assumptions about ghosts in the machine, central executives and other undescribed mediating functions.

In the next chapter we will see how the 'big' aspects of human cognition (reasoning, logic and so on) required another critical evolutionary bridge to complexity – one which, indeed, transformed the cognitive system. In all cases, though, the cognitive system works at amazing speed and productivity. Every thought or new idea, every new concept or conceptual scheme, is the equivalent of a new 'adaptation' or structure of existence, arising on a scale of seconds rather than generations. Unlike the relative slowness and irreversibility of phylogeny, and even physical development, dynamic cognitive systems can shift both quickly and reversibly in order to deal with rapid environmental change.

9
Social Intelligence

There is another evolutionary bridge to intelligent systems that I have barely mentioned yet. This is the strategy of forming social groups or communities. Of course, it is a phenomenon that has reached great complexity in humans, and we still aren't sure why. It is still commonly supposed that humans started to socialise and become more cooperative as a *result* of their bigger brains and the more complex cognitive systems they'd already evolved: the result, that is, of an intelligent agreement to just get along together. Such an account is now considered to be too simple. The reason for that is the question of what it was that led to such fortuitous pre-adaptations – that is, big brains, complex intelligent systems – in the first place. We really need to know that if we are to understand the system at any level. The alternative – and, now, more acceptable view – is that social cooperation first emerged in other species as an occasional behaviour pattern; it's advantages became manifest; and so, in certain circumstances, it became subject to natural selection. The more complex cognition that social cooperation demanded evolved with it. In this chapter I explore this possibility.

Social life as a strategy of adaptation is widespread among animals, even among yeasts and bacteria. It varies from occasional associations, through loose aggregates to the intense cooperative lifestyles of modern humans. These variations are worth exploring not only because of their intrinsic complexity, but also because they might explain why the cognitive systems of some groups – and especially humans – is so *distinctively* complex. Eventually, I hope to show that social life has, indeed, increased in complexity with evolution; that it has a connection with the evolving complexity of intelligent systems; that, in the end, it provided that final bridge to the cognitive complexity to be found in modern humans; and also how, at all stages, it is connected with environmental structure.

Social life in bacteria

Some sort of social aggregation has been observed in all animal groups, from the smallest to the largest, in almost all ecosystems. Extremes of environmental change seem to be the conditions in which it becomes advantageous: a kind of mutual assistance when the going gets tough. Even some bacteria and slime moulds form social groups in response to extreme conditions. For example, if nutrients become depleted, species of the soil organism *Myxobacteria* start to secrete a mutually attractant substance. In response, approximately 100,000 cells aggregate through chemotaxis to form a single 'fruiting body'. The wave of intercellular signaling demands tight genetic regulation that has been studied in some detail.[1] Extensive morphological and biochemical changes follow within the cells. Thick-walled spherical spores (sessile and resistant cells that can withstand starvation until food reappears) are formed from the rod-shaped cells, new proteins are synthesised, and a mound-shaped fruiting body is formed.

Rather similarly, members of the slime mould *Dictyostelium discoideum* roam around in the soil in an amoeboid form most of the time, foraging for bacteria to eat. When food becomes scarce they secrete a chemical signal that induces thousands of them to aggregate as a kind of 'slug' that moves towards light and heat, or otherwise more favourable microenvironments. Around 20% of the cells then form a long stalk that supports a spherical fruiting body upright on the soil surface. This produces showers of spores that attach to, and become dispersed by, passing organisms. Again, the changes in metabolism, cell adhesion and so on entail complex intercellular signaling systems and tightly regulated genetic recruitment.[2]

More active forms of social aggregation in single-celled organisms are also known. For example, the social bacterium *Myxococcus xanthus* appears to prey cooperatively on other bacteria. The bacteria feed in a swarm that permit them to overcome large prey that they cannot digest as separate cells. As with the chemotaxic and other stimuli, mentioned in Chapter 4, all these cases involve complex signaling systems picked up as spatiotemporally structured cues from the medium, regulating the onset of the new behaviours and metabolisms.[3]

It is tempting to describe the aggregate in such cases as one multicellular organism. Strictly speaking, though, they lack the true clonal nature and identical genetics of the cells of such organisms. The process, however, is a self-organising one that emerges from interactions among the individual cells, without a distinct executive or supervisory agent. Individual cells do not bear a genetic 'code' for the form of a fruiting body, nor for their individual roles in its formation. It is achieved by epigenetic and physiological responses to intercellular signaling systems, themselves responding to extreme environmental change. This brings about true cooperation in the literal sense of 'operating together'.

Taken together as a whole system – the cells embedded in the emergent social regulations – the cooperative strategy obviously reflects increased complexity, compared with non-social species, in the sense of numbers of variables interacting at various depths. They would soon, therefore, have become targets of natural selection. As the advantages of social aggregation emerged in evolution, the internal systems of individual cells, including their genetic resources, would have been selected for ability to participate, which would have enhanced aggregation and so on, in a co-evolutionary cycle. Indeed, it appears that some of the participants are more inclined to get together than others, and those that do tend to be more closely related genetically.[4]

Multicellular organisms

True cooperation is more obvious with the association of single cells to form multicellular organisms. Individual cells have become specialised and coordinated through epigenetic and physiological regulations, using highly structured signaling systems (described in Chapters 4 and 5). Emergent self-organising regulations must first coordinate the differentiation of the cells, and then the metabolic and anatomical divisions of labour. This is achieved through the structured signals they receive from within and outside the cell. The same sort of story can be told in relation to cooperating tissues of the body and the emergent physiological regulations between them. In such systems we can see that the complexity of the whole intelligent system is greater than the sum of its contributing parts.

When we turn to consider more complex animals with evolved cognitive systems, it is fairly clear that cognitive systems have evolved for the coordination of behaviour both in relation to the environment, generally, and in relation to social life. We need to try to pick out the different effects of these. We can ask, for example, how cognitive systems are different in *social* species compared with non-social species? As part of that question we can also ask whether the interactions between 'social' individuals are more complex than those between 'non-social' individuals, or more complex than interactions with inanimate objects? We can further explore the nature of any increased complexity, and what benefits it affords. Overall, though, we want to know what, exactly, has social life contributed to the complexity of cognitive systems, and *vice versa*? Let us look at the possibilities more closely.

Cognitive demands of social life

Behavioural coordination between individuals in a species is particularly well illustrated in ants, termites and bees. These species form highly regulated colonies, with distinct divisions of labour, involving one or a few reproductive pairs and many sterile castes. The latter are morphologically, physiologically and behaviorally specialised for different, mutually beneficial, roles. The benefits of this lifestyle appear to have been huge. It has been reported that ants form 10% of the animal biomass of the world, and termites another 10%, so that social insects (not mentioning bees) could make up a remarkable 20% of the earth's animal biomass.

The extreme phenotypic specialisation, and the strong social cohesion of these groups, arises from systems of developmental plasticity (another example of development itself adopted as an evolutionary strategy). A sugar-rich substance, 'royal jelly', is produced by workers, and contains the hormones that turn larvae into fertile queens, rather than sterile workers. One or more of these bloated queens subsequently produces all the eggs. These, in time, give rise to the variety of behaviourally specialised forms: workers, soldiers, nurses, winged nymphs that become reproductive adults and so on, depending on species. The environmental factors influencing this differentiation, as well as the proportions of types, are not well understood. In some cases, at least, they involve chemical pheromone signals secreted from mature members of castes (as if 'calling up' more of their own when their numbers are low).[5]

By using a variety of communicational devices between members, the colony can rapidly and coherently respond to the uncertainties of a changing environment: the availability of food sources; the presence of intruders; other physical disturbances; threats from predators; weather disruption and other factors.[6] The sum total of responses can be very impressive. Termites, for instance, build intricately structured nests with a complexity far beyond the comprehension of the individual termites. In their general behaviours, the crowds of individual ants become spatiotemporally structured by their own dynamics, determining 'the speed and plasticity at which the colony, as a whole, responds.'[7] The behaviour of some individuals, as in laying down pheromones leading to a food source, changes the environment to evoke new responses in others (reciprocal activity known as 'stigmergy'). But the joint structure of colony and environment exists only as the spatiotemporal parameters of interaction within the joint dynamics. It is not 'written' in a genetic or other kind of program.

Not surprisingly, then, insect colonies are often presented as prototypical cases of self-organising systems with emergent levels of regulation. Indeed, using a reverse logic, a number of computational scientists have adopted algorithms derived from observation of colony behaviour to solve computational problems. The interesting thing, as William Sulis notes, is that 'the low level of the individuals and the high levels of the collective have distinct descriptions and dynamics'.[8] Tadeusz Szuba reckons that these computations can only be achieved through a chaotic dynamical system.[9]

This possibility has been well researched. 'Ant societies, composed of interacting chaotic individuals, which can generate regular cycles in the activity of the colony, provide one of the very first examples of this phenomenon. They self-organize to attain nest densities at which the transfer of information, per capita activations, and the information capacity of the colonies, are maximal. At such densities, ant colonies are poised in the neighborhood of a chaos-order phase transition.'[10] Guy Theraulaz and his colleagues used an experimental set up to show how the spatial patterning of the way ants remove and pile dead bodies from the nest, has self-organising properties based on non-linear dynamics.[11] Likewise, the formation of nest clusters, formed as a predator defense, has dynamical properties.[12] Therefore, as well as the coalitions of chaotic attractors within individual brains, as mentioned in Chapter 7, such social life also consists of 'coalitions of coalitions' *between* brains.

In spite of that, it is debatable whether the demands of social life on the cognitive systems of individual social insects are much greater than the demands of the environment generally. Although their social interactions have an orderly, emergent construction, the environment for each individual still has a fairly stable structure. This is reflected, indeed, in the stereotyped reactions of the different castes and relationships between them. They interact on the basis of a limited range of individual behaviour patterns that have been constrained through canalised (closed), rather than cognitive (open), development.

In constructing the shortest path to a food source, for example, 'ants as agents have extremely simple rules of gradient-following and pheromone dropping.'[13]

These evolved social strategies, then, are a very effective way of dealing with environmental changeability within certain limits. But, as with the slime moulds, the patterns of social signals being responded to are not a lot different in complexity or depth than signals from the dynamical physical world generally. They are managed with limited changes to cognitive (or other internal) systems. Stephen Guerin and Daniel Kunkle refer to such systems as 'thin' agents, having 'simple rule bases', in contrast with 'fat' agents capable of 'internal reasoning' in more complex cognitive systems, to which we will turn below (see Note 13).

Social life gets more demanding

I mentioned earlier how the Cambrian explosion in varieties of species coincided with increasing intensities of predator–prey interactions, and thus improved hunting and defensive systems. One result may have been a kind of arms race. But another was the increasing evolution of social formations. Examples of that are seen today in numerous species of fish. Indeed, some researchers have suggested that such species exhibit complexity of social behaviours that were once thought to be the reserve of primates, even humans. 'Gone (or at least obsolete) is the image of fish as drudging and dim-witted pea-brains, driven largely by "instinct". Now, fish are regarded as steeped in social intelligence.'[14] This intelligence includes complex social structuring such as rank ordering of priority in mating and feeding; cooperating in hunting and predator defence; learning from the consequences of others' behaviours; and acquiring quasi-cultural traditions, as in repeated use of same nesting sites, and learned feeding and escape techniques.

The dynamics of fish social behaviour have been well studied. Shoals and schools exhibit complex formations for avoiding and confusing predators, while managing to avoid collisions between individuals or other obstacles, or breakdown in the coherence of the overall patterns. The very coherence, switches and variability of these patterns imply some organising agent. But, again, there are no executive or supervisory agents at work. All the group structural choreography is achieved by mechanisms within individuals obeying the structural dynamics emerging within the group as a whole.

Unfortunately, as with herds of ungulates and flocks of birds, the mechanisms for achieving such coherence are still poorly understood.

Experiments in fish tanks using individual tracking devices suggest that the structure being obeyed – the 'rules' of dispersion within the group – appear to change from group to group, from time to time.[15] Accordingly, individual cognitive systems will need to continually adjust to a switching structure.

Similar findings seem to apply to other social structures in groups of fish, such as the dominance hierarchy. Found in many species, such social differentiation regulates relative priority in terms of access to food, mates or other resources. Deterministic models have generally assumed that rank status in such a hierarchy reflects the level of some relevant individual attribute, such as size, aggressiveness or other indication of 'biological fitness'. But experiments by Ivan Chase and colleagues question that idea. They brought groups of fish into communion to form a hierarchy. The fish were then separated for a time before being brought back together again. In those reunions of exactly the same fish quite different hierarchies tended to form.[16] Although individual attributes did seem to have a minor role, social dynamics seemed to be much more important. This is confirmed by other studies in a wide range of species. They show that social hierarchies are self-organising, self-structuring, dynamic phenomena, perhaps with the function of reducing uncertainty in social access to resources.[17]

The real question, though, is what are the implications for cognitive systems and *their* complexity? It is generally assumed that social interaction at this level requires some 'extra' cognitive powers. But, there is uncertainty about what form these might take, based, in part on vagueness about the nature of the cognitive demands themselves. Nevertheless, a review by Ralph Adolphs, suggests that, 'Compared to the physical environment in general, the social environment is more complex, less predictable, and, critically, more responsive to one's own behavior.'[18] Some further insight into this claim may be gained from recent research on individuals' awareness of the 'minds' of other individuals in a wide range of species.

Cognitions about others' cognitions

There seems little doubt that animals' concepts of animate objects, generally, are likely to be more complex than their concepts of inanimate objects, even in non-social interactions. For example, the spatio-temporal structure captured in a rabbit's concept of a fox is likely to be much deeper than its concept of a tree. The difference is perhaps well illustrated in the cognitive deficiency of some humans with brain damage who can provide precise descriptions of non-living artefacts, but only impoverished description of living things. For example, a (human) victim of viral encephalitis, affecting the temporal lobe of the brain (an area particularly involved in recognition from concepts), was studied by Glyn Humphreys and Emer Forde. The patient could successfully describe a compass as 'tools for telling direction you are going' and a briefcase as 'a small case used by students to carry papers'. But when asked to describe a parrot he said 'don't know'; and he described a snail as 'an insect animal'.[19]

This difference seems fairly obvious. Other animals present more sets of articulated parts, all in spatiotemporal coordination, compared with most inanimate objects, and are usually themselves centres of creative action with variable structure. And, as Adolphs says, animals *respond* to other animals. Being able to form concepts of such experiences, and use them in making predictions, definitely implies some cognitive ability over and above that required for dealing with inanimate objects.

There have consequently been many claims about a kind of 'theory of mind' – or concepts of the 'rules' of others' behaviours, and their 'points of view' – even in sub-primate species. For example, Milind Watve showed how bee-eaters (birds) would not return to their nests if they saw a potential predator in the vicinity, but would be less cautious if the predator couldn't actually see the nest. On the other hand, they were very cautious if the predator had, in fact, already seen the nest before, even if it couldn't see it at that moment. Watve and colleagues argue that, in order to behave like this, the birds must have formed an impression of what the predator 'knows' about the nest – that is, have some knowledge of the other's knowledge.[20]

In another study, Nathan Emery and Nicola Clayton noticed that jays that had buried food for later retrieval (as they habitually do) when other jays were around, would tend to return later, alone, and hide the food in a different place. They were even more likely to do so if they had experience themselves of stealing other jays' food (as they also habitually do). This behaviour was confirmed in laboratory experiments, raising the possibility that jays are able to attribute mental states to other jays by constructing a complex model or 'theory' of the others' minds.[21]

These are not however abilities to do with social life, in the cooperative sense, so much as general (non-social) interactions with other animals. However, they may be indicative of ways in which social interactions could encourage the evolution of even *more* complex cognition. Indeed, there is evidence that social dominance hierarchies in birds do just that. One of the tests used to assess cognitive ability in animals has been whether they can solve transitive inference tasks. This involves making inferences from a logical order: for example, if Object A is bigger than object B that is bigger than object C, then object A must be bigger than object C. Such abilities of logical inference have been found in a range of mammals, but also birds, in social groups (and, of course, imply an emergence from reflective abstraction, as described in Chapter 8).

Paz-y-Miño and colleagues allowed stable dominance relationships to emerge in groups of captive jays. Individuals were then allowed to observe a dominant bird from their own group lose encounters with a bird from another group, while that latter bird was also seen to lose encounters with birds from a third group. After seeing such a fall from power, the observers from the first group responded to their own dominant bird less submissively. This suggests that they had retained some model of the structure of external relationships. In other words, they seemed to apply a transitive inference to their behaviour (in effect, that the 'alpha' bird was not as 'big' as they had previously conceived).[22] A number of related studies show how more complex social life seems to have demanded and produced more complex forms of transitive inferential abilities, as well as other cognition abilities.[23]

The idea that social life demands the ability to form a 'theory of mind' has been applied to other social vertebrates, especially primates. Observers regularly claim that monkeys and chimpanzees can deceive others, form alliances, predict what others can or cannot see, and generally make subtle inferences about the mental states of others. However, at least some of these claims are disputed.[24] It is, of course, difficult to know what 'theory' or 'model' an animal is carrying in its head. Nevertheless, the 'social intelligence' hypothesis argues that the large brains and improved cognitive systems of primates evolved as an adaptation to living in large, complex social groups.[25] Nicholas Humphreys claimed that the need for individuals to negotiate tricky rivalries and rank orders has demanded a new kind of smartness, and thus bigger brains.[26] A 'Machiavellian hypothesis' has been advanced based on observations of deception and cheating among members of a few primate groups.[27] By such cognitively 'clever' means, individuals are said to reap the benefits, while minimising the costs, of social life.

Such reports have invited the view of primate intelligence in general, and human intelligence in particular, as a process, not so much of cooperation, but of individual cleverness in manipulating social forces for personal ends. But it is difficult to substantiate the claim without assessing the complexity of the social life in more analytical, informational terms. I will be looking at this matter again later.

Imitation and 'culture'

One advantage of social life – and one that may be related to enhanced cognitive complexity – is the opportunities it provides to learn or acquire skills, such as tool use, from others. This is imitation learning and is widely observed in primates. For example, chimpanzees will start to use a stick to 'dip' for termites, or monkeys might start to wash potatoes on the seashore, after observing similar behaviours in peers. The claim is made that this amounts to 'cultural' transmission of knowledge and cognitive abilities across generations, similar to that in humans.[28] Andrew Whiten and his colleagues point to 39 distinct behaviours that they say are culturally transmitted.[29]

Closer studies of imitation confirm that it entails greater cognitive complexity than is usually supposed.[30] As described in Chapter 8, learning about objects and events in the physical world involves acquiring the deep spatiotemporal structures peculiar to them. Individuals generally must assimilate that structure through personal action upon those objects and events. Imitation is a kind of short cut through this abstractive process, the crucial structure being revealed/uncovered by the already-intelligent action of others. Personal behaviours can then, as it were, directly plug in to that structure and enjoy its fruits.

The novelty of each situation, though, means that the action is rather more than a simple motor 'copy' of another's responses. For example, the perspectives of ones own and others' behaviours are quite different. The individual needs to put him/herself in the place of the other and assimilate what the other sees and does to his/her own perspective. Then the visual impressions still have to be translated into motor actions. So the imitation still demands much cognitive transformation.[31]

On the other hand, precision of observations and reports by investigators in this area may not always be very high. Richard Byrne and colleagues warn that definitions of culture as applied to non-human primates have been vague and variable, and that a richer account of the cognitive contents of culture need to be developed.[32] Perhaps it is not surprising that Bjorklund and colleagues point to the 'often acrimonious debate about the specific nature of these abilities.'[33] Moreover, it has also been suggested that the perception of these abilities as 'special' in primates is more due to the tendency for them to be studied more intensively in small sample sizes over long time periods. The idea that primates face unique social and environmental challenges that make them more cognitively sophisticated than other mammals is itself now being challenged.[34]

It has even been suggested recently that imitation behaviour requires little if any conceptual processing at all. This follows discovery of the so-called mirror neurons in the cortex in primates (mentioned in Chapter 8). These neurons are activated by observing, or mentally rehearsing, an action, as well as by actual performance of the action.[35] Such properties are said to provide a direct conduit from observed behaviour to replicated motor patterns: that is, translation from a kind of action 'copy' in the mirror neurons to motor programme, without intervening cognitive analysis-synthesis.

However, there is still a lack of consensus about this idea. For example, some experiments have revealed activation of mirror neurons just by a reaching action.[36] Moreover, the mirror neuron hypothesis seems to ignore the complexity of cross-modal transformation of sensory inputs, from visual to kinaesthetic patterns, in the dynamic creation of the motor programme. The proposition that imitation needs some degree of complexity of cognition, over and above that used in normal interaction with objects, still seems reasonable.

A social brain?

If the demands of a more complex social life indeed required – and, therefore, fostered – a more complex cognitive system, we would expect some reflection in the brain size and structure. Teasing out the effects of social factors from those of other factors on brain size is, however, not straightforward. For example, brain size increases with body size, and is related to other aspects of habit and behaviour, such as diet. Carnivores generally have a size increment in frontal cortex compared with other mammals, thought to reflect the extra demands on cognition of a hunting lifestyle.[37] Primates, generally, have bigger brains through finer visual discrimination and digital dexterity. Those that forage for fruit have bigger brains than those that eat leaves, perhaps because they need a better memory for more scattered locations. And those that eat insects, have still bigger brains because they need still finer sensory discrimination and motor skills.[38]

In spite of that difficulty, correlations are reported between size of cortex and group size and group complexity in primates. That may reflect the additional cognitive competences required for living in a group, such as forming, and conceptually summarising, social networks. But the increment is not very great.[39] Moreover, there is no evidence to support this social brain hypothesis among carnivores that live in groups in comparison with those that live more solitary lives.[40] As Robin Dunbar says, the question is whether social situations are such as to require different or greater cognitive powers than everyday, 'ecological', problem-solving. 'Following the lead in developmental psychology, there is a growing view that social intelligence may not be a special module...but rather is a reflection of the ability to use basic executive functions in a more sophisticated way'.[41] Another review concurs that ecological factors may possibly have been more important than social factors in primate brain evolution.[42]

One thing does seems fairly certain, so far, then. If there *are* increments of brain size attributable to social life at this (pre-human) level, they are not huge (again, as compared with increments due to demands of the physical world, including non-social interactions with other animals). In any case, as Harry Jerison has said, much social complexity can be achieved without much increase in brain size.[43] More important than sheer size and (assumed) processing complexity may have been modulation of emotional and motivational expression by cortical centres.[44] Cognitively, however, it seems reasonable to suggest that the kinds of social life considered so far require abstraction of informational structure a little more complex than the demands of life in general, but not greatly so. Those demands would not seem to account, for example, for the tripling in brain size between humans and other primates, nor the superb cognitive abilities of humans. Perhaps we need to ask: is there some other factor?

Cooperation and epicognition

Perhaps the other factor may be the complexity of cooperation itself: it demands more complex cognition and, therefore, more brain network capacity. As we have seen, the social life of insects seems to be governed by relatively simple sets of signals. And schools of fishes, flocks of birds, herds of ungulates and so on follow physical rules that would appear to require little cognitive ability over and above the demands of the physical environment. In those cases the benefits of social life – defence, food foraging, mating – can be enjoyed with only rudimentary social regulations: following pheromone trails, or other simple cues, in social insects; a few simple rules in the case of vertebrate schools and herds; or sets of calls and gestures in the case of primates. There is little joint activity as such. True cooperation, on the other hand, requires a different level of complexity of cognitive engagement. In even simple tasks such as moving an obstacle together, each individual must be aware of the viewpoint of others, anticipate their actions and generate actions to coordinate with them.

Complex cooperative activity, as in hunting, foraging, defence and so on, is rare in pre-human primates. Although living rich social lives in other respects, monkeys and apes rarely help group members other than close family, and joint action and 'teaching' are also rare. There is little evidence among chimpanzees of agreement to share or of reciprocation.[45] For example, chimpanzees do not take advantage of opportunities to deliver food to other members of their group.[46]

Chimps alone seem to exhibit occasional cooperative hunting. Reported examples include raids on neighbouring groups for killing some of the males, and capturing some of the females; or forming alliances with other group members to overthrow a dominant male.[47] It is difficult to be sure from sporadic observations what the intentions, cognitions and motivations behind such rare behaviours really are, but they may well indicate rudimentary social cooperation. As such it suggests a kind of 'flicker' of true cooperation at this level of evolution.

Such sporadic cooperation may occur in other vertebrates, if reports of cooperative hunting are correct, even in some species of fish. In a review, Redouan Bshary and colleagues say, 'Co-operative hunting in the sense that several predators hunt the same prey simultaneously is widespread in fish, especially in mackerels, which have been described herding their prey.' It is observed that 'individuals play different roles during such hunts (splitting the school of prey, herding the prey) and refrain from single hunting attempts until the prey is in a favorable position.'[48] Presumably, this reflects individual cognitions 'cooperating' with a few simple signals that have emerged in the group, much as in social insects such as termites.

The cognitive demands of true cooperation, however, are very great. Acting jointly in a dynamic activity such as hunting requires a rapidly updating *shared* conceptualisation of the target, with high demands on attention. In addition, each individual needs to form momentary sub-conceptions, corresponding with their complementary place in the global perception. Finally, the coordination between global and individual perceptions must give rise to individual motor responses that are also globally coordinated. And all this rapidly changes in seconds or even milliseconds. This coordination could not happen without some 'epicognitive' regulations emerging among the individual cognitions of participants, by analogy with the epigenetic regulations coordinating signaling and recruitment of genes in cells.

Social cooperative hunting as a major survival strategy is prominent in a number of species of canids (wild dogs) and felines (wild cats). Among the social canids, cape hunting dogs have been well studied for their habit of running down prey through divisions of labour, surrounding the prey, strategic delays and other tactics of organisation. Similar patterns are observed in wolves and lions.

Again, such behaviour demands joint perceptions and actions, involving coordination of personal perceptions and actions embedded in more general (epicognitive) ones shared by the group as a whole. Such groups have been reported as having bigger brains, but not vastly different from corresponding non-social species.[49] However, there may have been an evolutionary limit to further brain expansion in quadrupeds because a horizontally extended neck would have been unable to bear the greater weight of a bigger brain in a bigger skull. This awaited the emergence of a bipedal (upright) gait, as described in the next chapter.

In short, true cooperation, as in cooperative hunting, may suggest the early foundations of an emergent 'epicognitive' level of regulation. This level arises from the joint contributions of participants, but then transcends them, before being refracted back through individuals in rapidly changing expressions. We have seen how coalitions of attractors operate within individual brains. Here we may be seeing a new level of abstraction from the coalition of brains *between* individuals. This suggests powerful dynamical systems, including an even more potent basis for reflective abstraction. That implies much for the complexity of cognitive systems, and the next chapter is about those implications.

10
Intelligent Humans

An impossible bridge?

Many theorists would disparage the notion of some kind of innovative leap in the course of human evolution. For example, Richard Byrne and his colleagues warn against a 'sterile searching for a cognitive Rubicon' between apes and humans.[1] Yet evolutionary leaps, or phase transitions, have occurred in the sense that major innovations, only nascent in a species at one time, have evolved in subsequent generations as a major adaptive strategy. The 'new' strategy has then accelerated evolution and species radiation. Such was the leap from prokaryotes to eukaryotes with the evolution of bounded nuclei; from single-cell to multicellular species; from invertebrates to vertebrates; from acquatic to terrestrial living and so on. In this book, I have described a series of such bridges in the evolution of complex, intelligent systems, each with remarkable consequences for further evolution.

In other words, in our Darwinian haste to emphasise the unity of species, we can sometimes neglect the differences, especially differences in complexity. When we turn to look at humans we can see how a number of such innovations make us very unusual apes. This is evident in bipedalism, manual dexterity, rich tool-making as a central, rather than casual, activity, and wider behavioural adaptability. Humans also have complex technologies, cooperative systems of production and a system of communication radically different from any other in the animal world. All other species – including apes – are specialised for a particular environmental niche; not so humans, who have no particular niche, colonising every corner of the world, and adapt the world to themselves rather than *vice versa*. All this reflects a cognitive system that is vastly more complex than that of even our nearest biological relatives, and seeming, to many people at least, to transcend evolutionary biological rules. More seriously, it still creates deep uncertainty among psychologists and cognitive scientists about what kind of system we actually have.

What really stands out anatomically, of course, is a huge increase in brain size, reflecting a three-fold expansion over that of other apes, even when account is taken of body size. And this expansion occurred over a relatively short evolutionary period. Nearly all of that difference lies in a vast neocortex, especially in frontal lobes, the areas most involved in coordination of activities, and integration of what we feel with what we think and do. This, in turn, seems to reflect the sheer number of brain cells and their interconnectivity in a 'space-saving' package.[2] Ten billion neurons and over 50 trillion connections are numbers frequently reported, also implying an exponential increase in their interactions. What could it have been about the environment and lifestyle of evolving humans that fostered such a remarkable change? And what does that tell us about the nature of the functions that have emerged?

Cooperation

Perhaps the first thing that a visitor from outer space would notice about this odd species, however, is the degree to which its members are engaged in cooperative activities. It is doubtful whether non-human primates engage in true cooperativity. Even some primatologists have been of the view that human intelligence is completely different, in that sense, from the intelligence of apes or monkeys. David Premack and Ann James Premack say that a major manifestation of this is an agreement to share, which again seems almost unique to humans. There is little evidence among chimpanzees of agreement to share, or of reciprocation.[3]

On the other hand, '[t]he happy tendency to share resources equitably – at least with members of one's own social group – is a central and unique feature of human social life'.[4] An important aspect of this characteristic is the sharing of knowledge: humans alone teach others of all ages. Young chimps, and juveniles of other species, can acquire various adult skills by self-learning or imitation, as mentioned in Chapter 9; they are never taught. On the contrary, teaching seems not just absolutely mandatory among humans, but also universally enjoyable.

More generally, humans cooperate in their survival and other activities in ways only vaguely prefigured in other species, including primates. Whereas providing for individual needs remains very much a question of individual activity for the chimpanzee, there is evidence even in the earliest human ancestors (hominids) of over a million years ago (mya) that humans cooperated far more intensively in activities such as group hunting and foraging, tool and shelter production, food preparation, defence, and distribution of resources. In the long run it seems to have produced a niche-busting plasticity of cognition and behaviour. Let us see how all this started and unfolded.

Human evolution

The circumstances that fostered the evolution of these abilities in Africa, some 7–8mya remain very hazy. Somewhere along the line there appears to have been another ratchet upwards of environmental complexity and uncertainty. The older story (up to the last decade or so) was quite neat. The climate dried, forests thinned and former forest dwellers were 'flung out' onto the open savannah or forest margins. Without natural defensive equipment, and deprived of traditional food resources, those species became extremely vulnerable. They had to face unreliable food supplies; requiring wide ranges and new diets; exposure to large predators and the like. They responded by cooperating with each other.

Some new adaptations were physical, such as the adoption of bipedalism, useful for seeing over tall grasses, and present in our earliest ancestors of 7.5mya.[5] But meeting the new demands for cooperation also presented further demands on cognition, far more complex than those experienced in the physical world alone, or in previous forms of social interaction. As mentioned in the previous chapter, such demands are present to some degree in the group living challenges experienced by most primates. But these were not great enough to make big changes to the brain and cognitive system, over and above those needed for dealing with more complex physical environments. Only in this new branch did it really make a difference. Defence, hunting and foraging became vastly more effective as an organised group, than as a mere collection of individuals. Similarly, while living in larger groups, reproductive relations, child rearing, divisions of labour, sharing of products and so on became less fraught when cooperatively ordered. The emergence of such patterns seems to have inspired a co-evolution of cooperation and cognitive systems: the demand for the former increased selection for the plasticity of the latter (and the self-organising dynamics that go with it) and so on. But there are other remarkable consequences that it is worth considering a little more closely.

Another evolutionary spiral

As mentioned earlier, some of that lifestyle is foreshadowed in cooperative hunters, such as wolves and cape hunting dogs. However, it was also foreshadowed by a suite of pre-adaptations already evolved in some form in the apes. Bipedalism occurs sporadically in some monkeys and apes to provide height of vision over tall grasses and shrubs, and wading in rivers, and was no doubt further selected in the hominid ancestors. This further freed hands for rudimentary tool use, which, together with the longer visual perspective of open spaces, would have improved visuomotor abilities. These factors may well have added the first increment in brain size, which would have been anatomically fostered by the new bipedalism and the ability to balance a heavier skull on top of the now vertical spine.

But it seems to have been the benefits gained from true social cooperation, communication and epicognition that inspired very rapid brain expansion. The immediate benefits started a virtuous evolutionary spiral in cognitive systems. Early hominids were gradually able to release themselves from the confines of a single habitat (to which even our nearest cousins are still tied). With deeper abstractions of the dynamics of the world, they were better able to anticipate and make change in their environments. By thus adapting the world to themselves, rather than vice versa, they were able to expand into many less hospitable environments.

The old story, though – or, at least its chronology – has been complicated by more recent fossil finds. As with any incursion of species into new habitats, it turns out that early hominid evolution was not so much a ladder as 'like a bush', with much mixing of features such as bipedalism, diet, manual dexterity and bigger brains.[6] A diverse array of related species, probably combining different permutations of these early features, and retentions of old ones, existed in various parts of Africa over the next several million years. Fossil findings have fueled many debates.

For example, mixed features of both tree dwellers and bipeds was evident 7–8mya, as is a dentition indicative of a more varied diet than the fruits and leaves of the strict forest-dwelling primates.[7] Other finds, dated from 4–5mya (*Ardipithecus ramidus* from Aramis, Ethiopia), shows a similar mixture of features. These include a braincase similar in volume to today's chimpanzees (340 cm^3 to 360 cm^3) and others indicative of life in thin forest.[8] Similar discoveries have suggested that the chief habitat of these creatures, in fact, consisted of open river systems, or lake margins, within otherwise crowded forests. Bipedalism, it has been suggested, would have evolved for wading in water.

From about 4.5 to 2mya, hominids are represented by a diverse group of undoubtedly bipedal, social living creatures with slightly bigger brains, around 450–500 cubic centimetres (cm^3). They had human-like teeth and hands. These have been called *Australopithecines* (the most famous being *A. afarensis* that existed between 3.9 and 3.0mya). Fossil dispositions suggest living in small groups of 20–30 individuals; and that lake margins and/or river floodplains may have been their habitat, with the availability of trees for escape from predators, especially at night.

These finds do not negate the general picture of evolution of cooperative lifestyle, described above, only that it may have been considerably more staggered in time. From about 2.5 to 1.6 mya, however, there appears, in the fossil record in East Africa remains of a variety of other species, with more human features, originally called *Homo habilis*. These are the earliest known members of the genus *Homo*; that is, the first 'human' species. They lived in a period of forest thinning when traditional forest fruits must have started to become scarce. *Habilis* sites are associated with a particular tool industry characterised by crude stone flakes, rounded hammer stones, and possible weapons regularly found near dismembered animal remains.

The disposition of remains suggests living in large groups, so they may have been social hunters, or possibly survived as scavengers. But the most important feature is an enlarged brain, in a cranium of around 650 cm^3, and therefore some 30% bigger than the Australopithecine brains, and much bigger than that of chimpanzees. In addition, they had lighter skeletons and skulls, with dentitions much reduced in size suggesting less reliance on tough vegetable matter in the diet. All this suggests some parallel between cooperative hunting and living in larger social groups, on the one hand, and between brain size and cognitive complexity, on the other.

A new species, *Homo erectus*, first appearing about 1.8 mya, is associated not only with another large increment in brain size but also with continuing brain expansion across the period of its existence: 850 cm^3 at 1.5 mya; 900 cm^3 at 800,000 ya; 1000–1200 cm^3 at 500,000–300,000 ya and 1200 cm^3 at 200,000 ya (modern *Homo sapiens* has an average brain size of 1400 cm^3). *Homo erectus* probably originated in Africa, but soon migrated across wide areas of Asia and Europe. They encountered some of the harshest climatic conditions on the way. This adaptability seems to have required closer cooperative life style, reflected in the brain expansions just mentioned, and in associated cognitive and practical abilities.

The most tangible of these in the fossil record is an increasingly sophisticated technology of stone tool manufacture. The uniformity of shape evident in tool manufacture also probably reflects a 'group' design, and thus a truly inter-cognitive culture of shared imagination and action. *Homo erectus* occupied sites indicating large social groups, and their hand axes and flints have been found around the remains of very large prey (elephant, bison and so on), suggesting advanced cooperative hunting skills. There is also abundant evidence at campsites of the use of fire, and perhaps cooking. It seems safe to assume that other cultural tools would have evolved, such as rules for group behaviour in such large groups, and communication more sophisticated than the thirty or so fixed signals of the chimpanzee.

Modern humans

Fully modern *Homo sapiens* emerged around 150,000 years ago, again probably 'out of Africa'. Fossils from that period indicate more sophisticated, adaptable and variable stone cultures, large social bands, with alliances among them, and extremely successful social hunting and home-making skills. The species also presents a huge increment of brain expansion, at the modern human average of 1400 cm^3: about three times bigger than the average for chimpanzees, even when body size is taken into account. As explained earlier, simple volume obscures a vast increase in folding of the cortex to increase functional tissue volume, and also changes in genetic regulatory and signaling networks.[9] The larger volume is also metabolically very demanding: human brains make up about 2% of our body weight, but use 15% of our oxygen, 25% of our metabolic energy and 40% of our blood glucose.

It seems reasonable to suggest that most of this enlargement is to support the more complex cognitive and epicognitive regulations needed for more intensive cooperative activity. Cooperative hunting on that scale would require advanced cognitive skills: intimate knowledge of the prey and their typical behaviour patterns; ecology; terrain; predator risks; weather and so on. But a new cognitive flexibility also seems to be reflected in the material culture of stone tools. *Homo erectus* and more recent 'pre-sapiens', such as the Neanderthals, were skilled toolmakers. But their stone culture, over a period of a million years, tended to conform to a stereotypical pattern or style. The period from about 70,000 years ago, however, is marked by a spectacular creativity and variability of styles, including complex bone technology, multiple-component missile heads, perforated sea-shell ornaments and complex, abstract, artistic designs.[10] This was probably also the case with other aspects of culture, the distributions of which suggest trading between groups. The main point is that the all-pervasiveness of 'cultural tools' in *H. sapiens* reflects true and obligatory epicognition.

As suggested above, this rapid emergence of a cultural lifestyle had co-evolutionary effects on cognitive systems. But it has made the biology of humans unique in other ways. Being equipped for dangerous living, widespread migration and constant testing of new habitats meant that our ancestors passed through some periods of intense biological selection. Such periods, which would have been ones of rapid elimination of the less fit, are referred to as ecological bottlenecks. As a result, humans are remarkably alike genetically. Although inhabiting almost every possible niche on earth, humans have only a quarter of the genetic variation of highly niche-specific chimpanzees, our nearest biological relatives.[11] Moreover, anatomical measurements show that the loss in genetic diversity has been mirrored by a loss in phenotypic variability in important characteristics.[12] In spite of that, human *behavioural* and *cognitive* diversity far outstrips that of any other species. Let us see how that complexity with diversity may have come about.

Complexity of epicognition

True cooperation is seen when even as few as two individuals act jointly over a common, non-routine task such as lifting an object – action extremely rare in all other species, but commonplace in humans. The partners in cooperative acts are obviously changing circumstances, not just passively adapting to them. But, in the process, they also jointly reveal a depth of forces that would not be experienced by either one operating alone. In jointly lifting an object, the natural relations between forces of mass, gravity, shape and friction, and the actions of one participant, all become conditioned by the actions of the other. Each has now to take account of that new complexity of forces in order for the joint action to become coordinated. In addition, of course, all these forces have to be geared to an overall, *shared* conception of joint purpose. In hunting and defensive actions against predators it is even more complex, because the object is itself active, and reactive, against the joint action. The dynamics of action and reaction aren't even remotely experienced by non-cooperating animals, and cannot be regulated through a narrow range of stereotyped chemical, gestural or other signals.

The complexity of such co-action, and its regulatory implications, have not been easily grasped within cognitive theory.[13] This is probably because the demands of such complexity on brain and cognitive functions has rarely been thoroughly analysed. Jerome Feldman noted how we underestimate the difficulties of coordinating multiple agencies, whether humans, animals or machines.[14] As Patricia Zukow-Goldring notes, ordinary experience involves a multitude of signals from a multitude of variables, through various senses, constituting what might be a multitude of objects (animate or inanimate) undergoing rapid spatio-temporal transformations. But in social cooperation, these inputs need to be coordinated between individuals so that 'perceiving and acting of one continuously and reflexively affects the perceiving and acting of the other.'[15]

Viewing the complexity in the context of dynamical systems, though, has been helpful. In social cooperation, one dynamic system interacts with other dynamic systems in productive and novel ways. The coalitions of attractor systems that operate *within* cognition, as mentioned in the previous two chapters, must now operate as 'epicognitive' sets of attractors *between* cognitive systems. As Frederick Abraham explains, social cooperation consists of two or more dynamical systems 'coupled', such that the state of one is a function of the state of the other.[16] Now imagine three, four or more of such systems coupled in, say, just a simple task like lifting an object. We get a non-linear dynamic system of non-linear dynamic systems – coupling 'intracognitive' through 'epicognitive' dynamics. A new level of self-organising regulations, with new emergent possibilities, has arisen. These can feedback and constrain individual cognitions in continuous non-linear loops.

Such processes are, of course, formidably difficult to observe *in situ*. However, computer and mathematical modelling of them has been encouraging. Computer modelling has used a variety of devices called cellular automata, multi-agent systems (MAS), or just 'artificial societies'. These consist of units, or 'agents', typically connected to each other on the computer in a 50 × 50 grid. Each unit represents an individual cognitive system, with given response tendencies (e.g., to go from an inactive to an active state given a certain input from other units). The units are then made to interact in response to some 'environmental' input, such as a signal along a particular connection. Keith Sawyer describes them as 'autonomous computational agents negotiating with each other in a distributed self-organizing fashion'.[17]

Such studies have demonstrated how the inter-agent dynamics enable the system to operate as a self-organising dynamical system. But it involves the emergence of an inter-agent level of regulations (i.e., a kind of 'epicognition').[18] In one study, James Crutchfield and colleagues introduced a 'model of objects that interact to produce new objects', creating a dynamical system that 'emerges on several distinct organisational scales during evolution, from individuals to nested levels of mutually self-sustaining interaction.'[19] The interactions can result in a diversity of dynamics including quasiperiodicity, stable limit-cycles, and deterministic chaos. Of course, as with ordinary cognitive systems, described in Chapter 8, this 'epicognitive' dynamics also produces a vastly expanded potential for reflective abstraction (i.e., creativity) – only now *between* brains as well as within them.

I will consider some of the possibilities and limitations of such a perspective, below. The point for the moment is that such interacting systems readily form a new level of dynamics with emergent regulations over the individual contributors. This is what seems to have happened with human cooperative groups. Such emergence does not constitute a special 'Rubicon', but the culmination of a trend for abstracting dynamic structure from complex, changeable worlds, that started at the origins of life itself. Human social cooperativity is simply vastly more complex in its demands on a cognitive system than the environments, social or otherwise, experienced by other species. It put the species straight onto another evolutionary escalator. It promoted the epicognitive dynamics, already peripheral in some earlier primates, to a central strategy of adaptability. The co-evolution of cognitive and epicognitive systems that followed probably explains a major part of the huge increase in brain size in human evolution.

Human culture – from evolution to history

Of course, there are numerous ways in which epicognitive regulations are manifested in human groups. Shared 'rules' emerge for coordinating joint actions. Examples include conventions such as turn-taking; familiar signals and gestures for demanding joint attention; marking aspects of what to attend to (e.g., pointing); means for indicating the end goal and route (strategies); plans for divisions of labour; signals for monitoring and modulating progress and, of course, in the use of a unique language. Just think of helping someone lift a long table through several doorways, or a wardrobe downstairs.

Other regulations are enshrined in shared mechanical devices and artifacts, from simple stone tools and shelters to emergent technologies, literally designed 'with others in mind'. Still others are the sets of overt, institutionally stated, obligations and conventions, as in marriage rules, kinship identities, asset ownership conventions and so on. These define rights and responsibilities, cementing patterns of relationships by legitimising them, often in stated laws. Human language, with its unique productivity and speed of transmission evolved specifically for coordinating social cooperation. Other sets of 'rules' were devised for simply celebrating and enjoying expressions of new-found structure, and the ability to discover it and create it. The shared appreciation of beauty – a property of the deeper structure in the natural and constructed world – and the joint creation of structure in art, music, dance and other expressions, all revel in that new ability. All these forms of regulation shape and cement the cooperative social relationships, most of which can vary radically from group to group or place to place.

The Russian psychologist Lev Vygotsky referred to these rules and devices as 'cultural tools'. They all reflect a depth and detail of informational structure not experienced in, or demanded of, previous species. Such social, epicognitive, constructions do not, of course, have a palpable existence in, as it were, the spaces between individuals. Just as regulations in the brain do not exist in individual neurons, but are distributed among the interconnections in a network, so the cultural constructions of humans are distributed in the mind-brains of group participants, as a dynamic coalition. Robert Wright has called this the 'invisible social brain'.[20] The dynamics of this distributed form has added a new dimension to cognition, defining the unique creativity of humans. This notion of culture is quite different from those theories conceiving culture in elemental forms, such as collections of symbols, memes and packets of information.[21]

I return to these aspects of culture later. Meanwhile, it is important to note that human social cooperation was not a voluntary process, in which a collection of 'brainy' individuals 'chose' to work together. Human cooperation was not 'invented' as a kind of clever idea, rather it emerged as a solution to stark environmental dynamics. In that way it became the context, not the result, of rapid cognitive evolution.[22] Humans attend to, and act in, the world as social conglomerates: their needs and aspirations are satisfied through social organisations, not individual adaptations. At its strongest, this view means that individual development is absolutely dependent on the acquisition of culture:

> ...[the human] nervous system does not merely enable [us] to acquire culture, it positively demands that [we] do so if it is going to function at all. Rather than culture acting only to supplement, develop and extend organically based capacities...it would seem to be ingredient to those capacities themselves. A cultureless human being would probably turn out to be not an intrinsically talented though unfulfilled ape, but a wholly mindless and consequently unworkable monstrosity. Like the cabbage it so much resembles, the Homo sapiens brain, having arisen within the framework of human culture, would not be viable outside of it.[23]

In what follows I want to emphasise several aspects of this all-pervasiveness: cultural tools as the means for coordinating individuals in groups; the consequences of that for cognitive processes; and its phenomenal creativity and adaptability. But I will be stressing their dependence on underlying structure as I go along.

Dynamics of cultural tools

The joint regulations necessary to coordinate social cooperation emerge dynamically among participants in the form of cultural tools. These include the vast range of rules, conventions, institutions, as well as technological tools fashioned for shared use, as we have seen. But they also become manifested as *psychological* tools through which individuals think and act with others. Accordingly, human cognition has become fashioned by its socio-cultural tools, just as the activities of neurons are fashioned by the global activity of their networks. As Lev Vygotsky put it, 'By being included in the process of behaviour, the psychological tool alters the entire flow and structure of mental functions. It does this by determining the structure of a new instrumental act just as a technical tool alters the process of natural adaptation by determining the form of labor operation.'[24]

Vygotsky concluded that human thought and activity is embedded in social life from the moment of birth. He argued that the entire course of a child's psychological development from infancy is achieved through social means, through the people surrounding it. The highly complex forms of human cognition can only be understood in the social and historical forms of human existence, he argued. One of the most quoted aspects of his theory is the 'law of cultural development': 'any function in the child's cultural development appears on stage twice, on two planes. First it appears on the social plane, then on the psychological, first among people as an interpsychical category and then within the child as an intrapsychical category'.[25]

This explains why human culture is not merely an efflorescence of human cognition, but its very medium and constitution. It also explains why cultural *differences* in cognition can be very deep. A large number of cognitive studies have revealed how people from different cultural backgrounds (say Asian versus European) use different 'cognitive styles', as well as varying in what they notice, attend to, remember, the criteria they use for object classification and so on. The social norms (or set of cultural tools) of North Americans seem to reflect a more individualistic, 'self-focused', cultural convention, compared with Chinese who are more aware of the connectedness of people in social contexts. Correspondingly, the cognitive functions also differ in aspects such as perceptions of figures, arithmetical tasks and aspects of language use (all cultural tools). Brain scans suggest that such differences seem to be reflected in differential activity in brain areas, as if the brain is being engaged in varied ways.[26]

Reflective abstraction in social context

There have been a number of efforts to locate, and amplify, Vygotsky's theory within a dynamic systems framework.[27] Many theorists expound dynamic views in a general sense, offering descriptions of development as the successive emergence of new cognitive structures.[28] The general message is that self-organising dynamics of lower level structures emerge as higher-order structures which, in turn, constrain the lower level dynamics.[29]

In the previous chapter I considered how attractor landscapes develop, first in more local attractor networks, then *between* networks as attractor coalitions. Reflective abstraction is an important product of the process. In social cooperation between brains, of course, we are getting synchronisation of these individual attractor landscapes at a still higher level. So these may give rise to deeper and more abstract reflective abstractions, and the logical and predictive abilities that go with it. These are, indeed, the epicognitive regulations that feed back into and help shape the individual cognitions. The important point is that the arrangement vastly expands perception and conception, furnishing uniquely human levels of consciousness and awareness of the world's affordances.

This enhanced perception is evident in so many aspects of human behaviour. Peter Carruthers notes the sophisticated grasp of fracture dynamics and the properties of stone materials in tool making, even among *Homo erectus* groups. 'Making stone tools isn't easy. It requires judgment, as well as considerable hand-eye co-ordination and upper-body strength. And since it uses a reductive technology (starting from a larger stone and reducing it to the required shape) it cannot be routinized in the way that (presumably) nest-building by weaver birds and dam-building by beavers can be. Stone knappers have to hold in mind the desired shape and plan two or more strikes ahead in order to work towards it using variable and unpredictable materials... Moreover, some of the very fine three-dimensional symmetries produced from about half-a-million years ago would almost certainly have required significant capacities for visual imagination – in particular, an ability to mentally rotate an image of the stone product which will result if a particular flake is struck off.'[30]

Such a richer and deeper conception of the world permits a 'metacognition', not just of the worldly structure, but also of oneself – a self-concept, identity and consciousness, perhaps only dimly available in apes. This includes awareness of the regulations through which we constrain our behaviour – that is, self-regulation. A weaver bird may create an intricate pattern in nest-building; humans can reflect on the pattern, talk about it, teach it to others, see possibilities for improvement, change it and so on. This acute consciousness means that humans can anticipate and predict events like never before. This new power of control over the world turns human evolution into human history.

Culture and biology

This account describes culture as a level of self-organising dynamics emergent from the dynamics of cooperating individuals. It is quite different from popular accounts of culture offered by theorists in evolutionary psychology, sociobiology and 'cultural biology'. These are based on attempts to understand human culture through the framework of traditional Darwinian theory, subject to the same selectionist logic as physical features. For example, they assume that what were originally individual learnings ('adaptations') are observed by others as solutions to *their* problems, and acquired by imitation (again as individual learnings). In the process 'errors' (i.e., 'mutations') may occur, increasing or decreasing the 'fitness' of what is learned, and making them subject to a kind of Darwinian natural selection, by analogy with genes. So the 'fittest' cultures – viewed as the aggregates of individual learnings – survive and spread.

I use the words 'kind of', and inverted commas, quite a bit, here, because the theorists themselves hedge quite a bit over how close an analogy to draw between culture and fixed adaptations.[31] Take for example, one ingredient, the proposal that the learnings shared can be treated as discrete 'units', like genes. In *The Selfish Gene*, Richard Dawkins called these 'memes' (from the Greek for what can be imitated). Although originally referring to things such as tools, clothing and techniques, the idea has been generalised to include ideas and beliefs, as well as specific behaviours. The analogy with genes, passing from person to person in biological inheritance, has obviously been very tempting. So culture is defined as pieces of information – ideas, beliefs, values, skills, attitudes and the like – acquired from other members of the species, affecting behavioural fitness, and therefore subject to natural selection.[32]

Perhaps not surprisingly, the theory involves a very narrow, individualistic definition of cooperation: 'Cooperation occurs when an individual incurs a cost in order to provide a benefit for another person or people'.[33] I hope it is now clear why these views seem erroneous. For example, all that has been discovered about genes and dynamic systems over the last few decades (see Chapters 4 and 5) militate even against *them* as units of biological currency in that elemental sense. Even at that level, functional units are actors in a more generally self-organised script. Moreover, these lower systems have been assumed into successively higher regulations to facilitate adaptability in rapidly changing environments. The complexity of such environments also confounds the idea of simple Darwinian selection in culture. It would require a near-constancy of conditions, for example. Rather, evolution has ultimately required a dynamic system that furnishes foresight, consciousness and predictability for changing the world rather than being passively subjected to it. Where is the Darwinian logic in a species that adapts the world to itself, rather than *vice versa*?

The attempt to understand one set of evolved regulations through the framework of another may explain why many people see the human species as 'a spectacular evolutionary anomaly' and 'the evolutionary system behind it as anomalous as well'.[34] I have argued that there is no anomaly, only a continuity of a largely overlooked evolutionary logic that has produced spectacular changes in complexities of living things. These consequences cannot be reduced to the logic of fixed adaptations in stable environments. Little wonder that, with the exception of peripheral cases, such as certain diseases or lactose-tolerance (the consequences of dairy farming), for which the particular environment is relatively constant, there is little evidence of natural selection in modern humans, especially at the level of cognitive systems. Therefore, this does not seem to be a fruitful way of looking at human culture.

A more dynamical perspective, both of environments and intelligent systems coping with them, yields quite different insights. Culture, as viewed through a dynamic systems framework, is not simply a veneer on individual mentality, but an emergent phenomenon that both absorbs cognition and is absorbed by it. Without that understanding the complexity of human mentality remains a mystery. To illustrate this, let us now look at how culture reflects on particular cognitive functions in humans.

Knowledge

Human knowledge, thanks to its cultural form, is far more insightful and deeper than knowledge in other animals. This is particularly well illustrated in human knowledge of objects. Because of our cultural connectedness, human use of objects virtually never consists of isolated cognitions around isolated entities. We know objects – their properties, uses and so on – through the patterns of social relationships of which they are an integral part. *Social* action with objects reveals far more than purely personal encounters possibly could, going much deeper than surface appearances and superficial physical properties. The socially embedded knowledge of objects transforms and extends individual cognitions, and, through a ferment of reflective abstraction, cultivates an intelligence about objects only vaguely foreshadowed in other species.

In learning about objects a child is also learning how they mediate social relationships, and they take on a far deeper meaning than inert physical entities. Children who – as in most school curricula – are required to commit to memory series of propositions about objects, materials and events, are only seeing the tip of an iceberg of the domain in question, and acquiring only a superficial understanding of it. Conversely, the more detailed properties of objects revealed by cooperative actions around them widens consciousness of them, and affords tremendous creativity. This is what probably suggested their appropriation as in tool use, tool manufacture, defence, shelter, warmth and so on. The abstract, aesthetic qualities of tools, and of the skilled actions of their use, also reflects the deeper structure of the environment on which they are used.

All this is reflected in the cognitive organisation of objects as object concepts (described in Chapter 8). In animals generally, this organisation remains useful to the individual for rendering further predictabilities about object properties. In humans, though, much more happens. Social cooperation requires us to be jointly explicit about our knowledge, as in sharing it, teaching it, working together with it and communicating about it. This means turning knowledge into a more 'declarative' or 'propositional' form through human language, in a way that cannot happen in other primates. In order to share our experience of an object with others, it must be ascribed to an agreed category, defined by a word ('table', 'furniture', 'artifact' and so on). Through the discourse tools of speech and writing, these more explicit knowledge structures can then be publicly reworked and extended into more complex forms. They can be further interrelated, used as metaphors for exploring new domains and so on.

In fostering more abstract conceptual schemes, then, social regulations radically alter cognitive regulations, a process that unleashes enormous cognitive powers in people. As concepts become extended in vast knowledge networks, predictabilities about the world become much deeper, and far more potent, than any found in non-human animals.

Science

This process has reached its most sophisticated form in the cultural tool we call science. Science is an institutionalised set of social processes for affirming shared knowledge. The implicit knowledge found in all other species is, of necessity, made explicit in humans when we need to communicate about it. Deeper layers of structure are exposed in the process. It has been the task of scientific research, from the times of Ancient Greece, to regularise that process. The main purpose is to discover and describe, in as precise terms as possible, the underlying structure of the natural world, as distinct from its surface appearances. Even prior to the more recent scientific revolution, this process radically altered some of our interpretations of everyday observations, such as whether the earth is really flat, or whether the sun goes around the earth or *vice versa*. In this way, predictabilities of future states from current conditions become ever more precise, permitting more intelligent interventions.

The socially evolved scientific method is specially designed for exposing those deeper structures in nature with clear definition. It is now well documented how the great flowering of science from the 17th century onwards entailed new knowledge-sharing procedures: making observations more systematically and repetitively; constructing explicit theoretical models; making predictions (hypotheses) from them; and testing these in controlled experiments. The need for all steps to be made explicit, so that they can be replicated by others, introduced the logical structure of modern empirical science. Indeed, it has been argued that the upsurge in the *social* processes of conferencing, cooperation and publication among scientists in the 17th-18th centuries was the single most important step in the historical development of the scientific method.

The current methods of science, that is, grew naturally out of epicognitive dynamics and perfectly illustrate the fruits of a cultural tool. But the rational social procedures thus historically evolved are also internalised by individuals as a personal psychological tool in cognition, permitting us to think scientifically as individuals. This is another example of the transformation and extension of otherwise limited cognitive abilities by cultural tools. It has, of course, been much assisted by other cultural tools. In particular, mathematics was developed precisely as a language for expressing the deeper structures and patterns in phenomena that cannot be expressed very easily in ordinary language.

The cultural institution of science is also the perfect example of joint reflective abstraction, reaching increasingly higher levels of abstraction, and vastly extending our knowledge-gaining capacities. Other organisms survive in changing worlds by assimilating the depth of structure therein. In bacteria this may be simple associations, perhaps conditioned by one or more other variables. The depth of structure increases across phyla. In humans, however, using various cultural devices such as mathematics and computers, it becomes theoretically unlimited. Thus we have managed to discover, and now talk about, non-linear dynamics, chaotic processes, and so on that remain hidden to other species. It is the revelation of hidden structure, in all its harmonious composition, in scientific models of reality, that led Albert Einstein to declare that 'there is nothing more beautiful than a good theory'.

Human memory

In pre-human cognitive systems, memory resides in the developed synaptic configurations within brain networks. In humans, memory also exists in epicognitive forms with the aid of cultural tools. Even since pre-literate societies, these have been expressed in verbal and graphic forms (drawings), and in song, dance, story and legend. At some time in human history, though, memories started to be shared as written symbols. Even the rudimentary forms of these new tools, such as marks on sticks, and, later, pictograms, furnished a crucial medium of social cooperation, but also of extended memory. As Vygotsky and Luria pointed out, the written forms of communication vastly expanded the natural memory function, transforming it into a new medium of cognitive organisation and planning.[35] Such auxiliary memory tools have, of course, been vastly augmented in more recent times by printing, libraries, calculators, computers and the internet.

Thinking

In Chapter 9, I offered a brief description of thinking as the self-organised activity within and between attractors, accommodating perturbations and translating predictions into actions. Even in non-human systems, thinking is often highly creative in dealing with ever-novel environments. But it remains harnessed to individual experience. In humans, however, thinking has become nested within an additional dynamics going on *between* individual cognitive systems. This dialectics of thinking has often baffled researchers who have sought to reduce human thinking to mechanical, 'logical' processes 'in the head'. For example, the logical abilities of individuals have been questioned when they have difficulty with problems such as:

All A's are B's
Are all B's A's?

Yet the difficulties evaporate when essentially the same problems are presented in socially relevant contexts, such as:

All humans are mammals
Are all mammals humans?

Indeed, cross-cultural studies have shown how judging what is or is not logical thinking is not a simple, objective process; rather the logic comes in socially constrained (epicognitive) forms. For example, Sylvia Scribner tried to administer logical syllogisms like the following to Kpelle farmers living in Nigeria, duly couched in local cultural terms:

All Kpelle men are rice farmers
Mr. Smith is not a rice farmer
Is he a Kpelle man?

But the first subject just kept replying that he didn't know Mr. Smith, so how could he answer such a question?[36]

Scribner had already shown that Western schoolchildren were better at answering such questions in the expected 'logical' manner? So are these Africans simply lacking in the logical abilities of Western schoolchildren? This turns out to be a premature conclusion. As Philip Johnson-Laird has pointed out, the subjects in such examples are actually making perfectly logical deductions from premises, for example:

All the deductions that I can make are about individuals I know
I do not know Mr. Smith
Therefore I cannot make a deduction about Mr. Smith.[37]

The difference lies in the dynamics of different, *socially* devised, logical forms. As Scribner points out, the standard syllogism depends upon another cultural tool of schooled children: that of using only information provided by the tester, and internal to the problem. This is typical of school-type tasks in which pupils are disciplined to repeat foregone conclusions from fixed premises. It reflects, not so much a superior logic, as a specific cultural device, entailing specific assumptions about what personal knowledge is to be 'activated' – a kind of game with tacit rules requiring the subject to ignore information outside of the premises given. Non-schooled people, on the other hand, tend to bring in *all* their background knowledge, as well as that of current context, in reaching an answer that is at least as logical. Whether or not someone is assessed as having logical ability, therefore, turns out, again, to depend on translation of cultural regulations into psychological tools.

Modulating feelings

The importance of affective variables, or emotions, in global attractor activities was mentioned in Chapter 8, with regard to individual cognition. As may be imagined, the need to integrate individual feelings, motivations and emotions into *joint* activities is a vital pre-condition of cooperation. In human psychology, so-called *emotional intelligence* is viewed as the assertion of reason over feelings, for purposes of delaying action or diffusing pressures, under social constraints. We would, therefore, expect epicognitive regulations to have means of modulating individual feelings. In other mammals, the frontal lobes of the brain appear to have a major role in such functions through rich interconnections with subcortical centers of emotion, as well as with all other regions of the cortex. The frontal lobes, and their connections, have undergone vast expansion in size in the course of human evolution.[38]

In more recent studies, John Allman and his colleagues refer to the discovery of a distinctive cell type, the spindle neuron of the anterior cingulate cortex (an 'interface' between cognition and emotion), the numbers of which have increased in primates from orangutans to humans. But they are especially numerous in humans, and their connectivity suggests that they function to coordinate centres involved in self-regulation of motivation, self-control and action planning. The authors suggest that, 'These neurobehavioral specializations are crucial aspects of intelligence as defined as the capacity to make adaptive responses to changing conditions', and further hypothesise 'that these specializations facilitated the evolution of the unique capacity for the intergenerational transfer of the food and information characteristic of human extended families.'[39]

This uniquely human form of intersubjectivity means that, just as we can fairly well think what others are thinking, we can *feel* what they are feeling, too.[40] Indeed, there is evidence from brain scans of joint activation of emotional centres in brain during mutual cooperation.[41] This aspect of our heritage has widespread consequences for human social life. These include an emotional underpinning to shared rules of social life; new abilities for delaying 'impulsive' activities in the interests of long-term planning; and a new source of motivation for the care of children. They have also brought huge enjoyments from shared constructive activity, often just for its own sake, as in music, dance and so on.

I suggest that this 'inter-emotion' is the true source of the unique degree of altruism found in humans. We can feel what others are feeling, just as we can think what they are thinking, and act accordingly. In pre-human social primates there is 'emotional contagion' in which an emotional state in one individual triggers the same state in another, in turn inducing some appropriate (e.g., predator avoidance) behaviour. But this is primarily individualistic – an aggregate rather than a sharing of feelings – only secondarily beneficial to others, without direct intention. There may have been evolutionary flickers of true inter-emotion in some social primates, but not as the major adaptation found in humans.[42]

This account of altruism contrasts with those found in the sociobiology and cultural biology literature. The latter tend to explain altruism in humans, as well as other animals, in terms of shared genes rather than shared minds. The general idea is that individuals will act for another's benefit if, in the process, it is likely to lead to perpetuation of their own genes, suggesting that apparent altruism is really only the 'behaviour' of selfish genes in disguise. There is now a huge, and controversial, literature on this subject that I won't take us into, here.[43] I see it as another attempt to reduce evolution to a single Darwinian logic, overlooking the dynamics of later stages of evolution.

The interlocking of epicognition with individual cognition and feeling, in fact, ensures that humans are obligatory altruists, as much as they are obligatory participants in all other aspects of their culture. But it makes us far more distinctively human in other ways. The whole developed attractor landscape lying between culture, cognition and feelings is probably what constitutes individual personality, with all kinds of implications for understanding human behaviour within a dynamic systems perspective. The complex also engenders feelings, and quests for, structure that we sense as wholeness, harmony and beauty.

Language

The unique, deeply structured epicognition in humans has required a very special form of communication. In structure, creativity and speed of operation, human language far exceeds the communication systems of all other species. Theorists have debated for centuries, right up to current times, where these properties come from. As so often when the nature of some function appears to be inexplicably complex it becomes attributed to the genes (usually without asking how the genes manage to create such complexity). Models of 'innate' language structures were proposed in the 1950's by Noam Chomsky. More recently, Steven Pinker has spoken of the 'language instinct', and insists that the capacity for language is rooted in human biology, not human culture.[44] Intensive searches are taking place for the special brain centres, and their associated genes, that have the technical properties for discharging the functions of language.

This is not the place to enter into this debate in detail. What is worth challenging, here, is the whole mode of analysis that views language as a distinct computational system or module in the brain. I suspect it reflects an inadequate analysis of the structure of natural and social experience and, therefore, of cognition. But there is an alternative view, which is that language is another cultural tool for mediating epicognition. It does this by translating the dynamic structures of shared perceptions and intentions into more linear, temporal patterns of sound and visual symbols: that is, in a form that can be transmitted to, and interpreted by, others. The syntactic and semantic structures of language reflect those of the dynamic epicognition it communicates.

Here again, though, we can see spectacular co-evolution of social function and individual anatomy. As language evolved as a cultural tool in the earliest cooperating hominids, so it became the context for changes to anatomical and neural networks requirements. The anatomical modifications include those to the shape and musculature of the jaw, mouth, pharynx and larynx. The brain networks probably involved nothing original or 'special', and no dedicated modules. Rather, they probably reflect an augmentation of those already used in apes for the analysis of finely graded sensory inputs and the construction of equally finely graded motor outputs, as in finger coordination and so on.[45]

Human language does, however, have considerable depth of organisation. Of necessity, its rule-structure must reflect the structure of cognitions as translated into more sequential format in a sound stream. Without such function it could not mediate the thoughts and feelings of people. Interpreting a received stream of speech depends on the detection of changes in numerous variables such as sound frequency, timing of voice and place of articulation of the tongue, as little as a hundred or so milliseconds apart. These discriminations are themselves embedded in (conditioned by) a complex system of variables for fuller semantic interpretation. And these are embedded in the epicognitive 'rules' of the culture, including conventions of language use.

The dynamics of the whole again amplify and extend individual cognition. Jean Piaget noted how the development of language in children 'enables the sensorimotor intelligence to extend itself.'[46] Vygotsky and Luria claimed that, once children learn to speak, their behaviour becomes entirely different from that of other animals. They noted how the development of language seems to impart to the child greater freedom of operation in tasks – a detachment from the consuming predominace of the goal itself to allow careful planning. This includes the preliminary gathering of a wider range of useful materials for tools, and the organisation of a chain of preliminary acts leading up to the goal. The impulsive, direct manipulations of the ape are replaced, in the child, by the complex psychological processes of speech-planning and controlled motor organisation. 'These entirely new psychological structures are absent in apes in even moderately complex forms'.[47]

In other words, as with other cultural tools, this socialised speech is progressively internalised as a *psychological* tool. By manipulation of its structure (as in thinking verbally), individuals can use speech to play a specific organising part in thought and action, often giving rise to novel ideas. As Vygotsky explained, we often come to solve a practical task with the help, not only of eyes and hands, but also of speech.[48] More generally, the individual is freed of slavishness to the immediate situation and becomes an active agent in foreseeing and planning social operations.

Language development also illustrates the important point that children must not be viewed as minds being passively programmed by social forces. Rather development lies in the 'collision' (as Vygotsky put it) between mature cultural forms of cognition with the less-developed forms in the child. The need to resolve such conflict means that the cultural order is, in a sense, reconstructed in each developing child, but often with novel individual forms that can feed forward to broader cultural change. These dynamics are the basis of human creativity, and also explain the vast diversity of human knowledge and thought, between individuals, between groups and between historical periods.

Above and beyond chaos

It might be tempting, from the above account, to view human social life as the natural, unconscious, working out of non-linear dynamic systems. Many aspects of human social life might suggest such a picture. For example, social level (epicognitive) regulations and their cognitive counterparts in individuals can display either long periods of stability, or of sudden transitions, that can duly be described in terms of the interactions within attractor systems. So theorists have drawn dynamical parallels between seemingly disparate social and biological phenomena: for example, between social revolutions and viral epidemics, or between international arms races and animal ecosystems.[49] 'Societal change in political and economic ideology can also occur in a rapid, nonlinear manner...with a trajectory that resembles phase transitions in physical systems'.[50] Likewise, cities have been modeled as sets of spatial chaotic attractors, as have public opinion, wars between hostile nations, international economic developments and social revolutions. It is said that even voting in elections sometimes exhibits chaotic dynamics due to interactions between individuals and voting rules.[51]

Does this suggest that human affairs are merely the subterranean operation of dynamic processes in which human consciousness is merely incidental, and self-determination an illusion? As suggestive as those ideas are, we need to be cautious. As Till Grüne-Yanoff warns, these theoretical developments may be overreaching themselves.[52] However fascinating NLD systems approaches may be for understanding the orderly behaviour of social insects, humans are not ants. As Robin Vallacher and Andrzej Nowak say, human groups 'are more than self-organised ensembles of simple particles. People have values and beliefs, moments of self-reflection and sudden impulse, universal concerns and idiosyncratic tendencies'.[53] More importantly, humans have the power to reflect on their 'operating rules' and to attempt 'to override them' – that is, metacognition. This reflects Vygotsky's point about those dynamics involving a 'conflict' between individual and social levels, with constant cross-fertilisation between them.

In a similar vein, Keith Sawyer argues that dynamic models based on simpler systems have tended not to incorporate the unique properties of human social systems. He argues that human social systems are qualitatively different because of the complex properties of human communication and language. Accordingly, 'chaos theory is not useful to social scientists other than as a source of provocative metaphors' (p. 25).[54] Most interestingly, too, Diane Richards points out how findings about the dynamics of elections (mentioned above) have implications for systems of democracy – implying that we can do something about them in a detached, conscious manner.

Has human socio-cognition, then, evolved, beyond passive subjection to dynamic forces to a level of conscious control over them? Actually within an NLD systems view there's no problem with such a proposition. Such consciousness may emerge naturally from the epicognitive dynamics through reflective abstraction. So hugely enriched are conceptions of the world created by epicognitive dynamics – especially in the form of empirical science – that we gain insights into the dynamics themselves, such as to intervene in, or work with, them. This is consonant with Vygotsky's observations, as mentioned above, that, when even young children acquire cultural tools such as language they also seem to obtain a detachment from experience that fosters new powers of organisation and planning. What emerges on the social plane, as a cultural tool, is reproduced on the individual level as a psychological tool – in this case as metacognition about the forces regulating our own cognition. I mentioned this process of internalisation, above, in all the examples considered – memory systems, logic, emotional intelligence and so on. So it is with human consciousness and our relations with the dynamics of the world. Unlike the passive, unplanned, dynamics of termite nests and cities, we can plan and organise ours from a quite different dynamical vantage point. This is the great, self-constructing quality of human intelligence that brought us from the plains of Africa to our modern insights and abilities in the world.

This, then, seems a satisfying culmination of evolution of intelligent systems – and perhaps of this book! In it I have tried to dispel the prevailing sense of mystery surrounding the evolution of cognitive systems, and the unfathomable complexity of the human mind, from its very roots in the complexity of life itself. It started as a dynamic living system that could persist in a dynamic world, through sensitivity and responsiveness to environmental structure. Through encounters with increasingly complex environments it evolved through several nested layers of structure-based dynamics. And then finally emerged as an offspring with a rational consciousness of the parent that gave it life. How we treat that parent, each other, and its other offspring, may become the ultimate test of our human intelligence system.

Notes

1 Why so complex?

1. Ayala, F. (2007) 'Darwin's greatest discovery: Design without a designer', *Proceedings of the National Academy of Sciences*, 104, Suppl. 1, 8567–73.
2. Darwin, C. (1859/1972) *On the Origin of the Species*, Harmondsworth: Penguin. p. 459.
3. Darwin, C. (1871) *The Descent of Man, and Selection in Relation to Sex*, London: John Murray.
4. Darwin (1859/1972) *On the Origin of the Species*, Harmondsworth: Penguin.
5. Heylighen, F. (1999) 'The growth of structural and functional complexity', in Heylighen, F., Bollen, J. and Riegler, A. (eds) *The Evolution of Complexity*, Dordrecht, Kluwer Academic; Adami, C., Ofria, C. and Collier, T.C. (2001) 'Evolution of Biological Complexity', *Proceedings of the National Academy of Sciences*, 97, 4463–4468; Miconi, T. (2008) 'Evolution and complexity: The double-edged sword', *Artificial Life*, 14, 325–44.
6. Dawkins, R. (1986) *The Extended Phenotype*, Harmondsworth: Penguin, p. 218; Dawkins, R. (1986) *The Blind Watchmaker*, Harmondsworth: Penguin, p. 222.
7. Miconi, T. (2008) 'Evolution and complexity: The double-edged sword', *Artificial Life*, 14, 325–44.
8. Salthe, S.N. (2008) 'Natural selection in relation to complexity', *Artificial Life*, 14, 363–74. p. 363.
9. Bird, R.J. (2003) *Chaos and Life: Complexity and Order in Evolution and Thought*, New York, Columbia University Press.
10. Gershenson, C. and Lenaerts, T. (2008) 'Evolution of Complexity', *Artificial Life*, 14, 241–3.
11. For example, Mazur, S. ed. (2008) *The Altenberg 16: An Exposé Of The Evolution Industry*, New York: Random House (a report of a meeting of sixteen prominent evolutionists and philosophers promoting an Extended Evolutionary Synthesis); Sterelny, K. (2002) *Dawkins vs. Gould: Survival of the Fittest*, Cambridge: Icon Books. Weisss, K.M. and Buchanan, A.V. (2004) *Genetics and the Logic of Evolution*, London: Wiley; Muller, G.B. (2007) 'Evo-devo: Extending the evolutionary synthesis', *Nature Reviews, Genetics*, 8, 943–8.
12. Greener, M. (2007) 'Taking on creationism. Which arguments and evidence counter pseudoscience?' *EMBO Reports*, 12, 1107–9.
13. Sinha, C. (2006) 'Epigenetics, semiotics and the mysteries of the organism', *Biological Theory*, 1, 112–15.
14. Richerson P.J., and Boyd, R. (2000) 'Climate, culture and the evolution of cognition', in Heyes, C., and Huber, L. (eds) *Evolution of Cognition*, Cambridge, MA: MIT Press. p. 337.

15. Boyd, R. and Richerson, P.J. (2005) *The Origin and Evolution of Culture*, Oxford: Oxford University Press. p. 11.
16. Shiffrin, R.M. (2003) 'Modelling perception and memory', *Cognitive Science*, 27, 341–78. p. 375.
17. Fodor, J. (2000) *The Mind Doesn't Work That Way: The Scope and Limits of Computational Psychology*. Cambridge, MA: MIT Press.
18. Dultz, R. (2008) 'A paucity of philosophy' (Letters), *The Psychologist*, 21, 1070–1.
19. Pinker, S. (1997) *How the Mind Works*, London: Penguin. p. 21.
20. Pothos, E.M. (2003) 'The rules versus similarity distinction', *Behavioral and Brain Sciences*, 28, 1–49. p. 26.
21. Editorial (2008) 'Bountiful Noise', *Nature*, 453, p. 134.
22. Mitchell, M. (1999) 'Can evolution explain how the mind works? A review of the evolutionary psychology debates', *Complexity*, 3, 17–24. p. 21.
23. Cosmides, L. and Tooby, J. (1994) 'Origins of domain-specificity: Evolution of functional organization', in Hirschfeld, L.A., and Gelman, S.A. (eds) *Mapping the Mind: Domain Specificity in Cognition and Culture*, Cambridge: Cambridge University Press. p. 96.
24. Pinker, S. (1997) *How the Mind Works*, London: Penguin. p. 21.
25. Gangestad, S.W. and Simpson, J.A. (eds) (2007) *The Evolution of Mind: Fundamental Questions and Controversies*, New York: Guildford Press.
26. Parter, M., Kashtan, N. and Alon, U. (2008) 'Facilitated variation: How evolution learns from past environments to generalize to new environments', *Public Library of Science: Computational Biology*, 4, e1000206; see also Kashtan, N., Noor, E. and Alon, U. (2007) 'Varying environments can speed up evolution', *Proceedings of the National Academy of Sciences*, 104, 13711–16.
27. Clark, A. and Thornton, C. (1997) 'Trading spaces: Computation, representations and the limits of uninformed learning', *Behavioral and Brain Sciences*, 20, 57–92.
28. Godfrey-Smith, P. (2000) 'Environmental complexity and the evolution of cognition', in R. Sternberg and J. Kaufman (eds) *The Evolution of Intelligence*, Hove: Erlbaum.
29. Godfrey-Smith, P. (2000) 'Environmental complexity and the evolution of cognition', in R. Sternberg and J. Kaufman (eds) *The Evolution of Intelligence*, Hove: Erlbaum. p. 10.
30. Richerson, P.J., and Boyd, R. (2000) 'Climate, culture and the evolution of cognition', in Heyes, C., and Huber, L. (eds) *Evolution of Cognition*, Cambridge, MA: MIT Press. p. 337.
31. Sporns, O., and Kötter, R. (2004) 'Motifs in brain networks', *Public Library of Science: Biology*, 2, 1910–18. p. 1910.
32. Flinn, M.V, Geary, D.C. and Ward, C.V. (2005) 'Ecological dominance, social competition, and coalitionary arms races: Why humans evolved extraordinary intelligence', *Evolution and Human Behavior*, 26, 10–46. p. 10.
33. Oyama, S. (2004) 'Locating development, locating developmental systems', in Scholnick, E.K., Nelson, K., Gelman, S.A. and Miller, P.H. (eds) *Conceptual Development: Piaget's Legacy*, Hillsdale, NJ: Erlbaum. p. 199.
34. Pigliucci, M. (2002) 'Buffer zone', *Nature*, 417, 598–9.

35. Dent-Read, C. and Zukow-Goldring, P. (2002) 'Where does the animal end and the environment begin...and end?', in Dent-Read, C. and Zukow-Goldring, P. (eds) *Evolving Explanations of Development*, Washington, DC: American Psychological Association. p. 552. See also, Lickliter, R. (2008) 'Developmental dynamics: The new view from the life sciences', in Fogel, A., King, B.J. and Shanker, S. (eds) *Human Development in the Twenty-First Century: Visionary Ideas from Systems Scientists*, Cambridge: Cambridge University Press.
36. Love, A.C. (2007) 'Reflections on the middle stages of evo-devo', *Biological Theory* 1, 94–7. p. 95.
37. Ahouse, J.C. and Berwick, R.C. (1998) 'Darwin on the mind', *Boston Review*, 23, 36–41. p. 36.
38. See note 36.
39. Edelman, G.M. (2006) 'Synthetic neural modeling and brain-based device', *Biological Theory*, 1, pp. 1–12. p. 8.
40. Maturana, H.R. and Varela, F.J. (1980) *Autopoiesis and Cognition*, Boston, MA: Reidel.
41. Maturana, H.R. and Varela, F.J. (1992) *Tree of Knowledge*, London: Shambhala. p. 75.
42. Lovelock, J.E. (1979) *Gaia: A New Look at Life on Earth*, Oxford: Oxford University Press.
43. Lovelock, J.E. (1979) *Gaia: A New Look at Life on Earth*, Oxford: Oxford University Press. p. 60.
44. Wagner, G.P., Pavlicev, M. and Cheverud, J.M. (2007) 'The road to modularity', *Nature Reviews, Genetics*, 8, 921–7; Kashtan, N. and Alon, U. (2005) 'Spontaneous evolution of modularity and network motifs', *Proceedings of the National Academy of Sciences*, 102, 13773–8.
45. Quoted in Callebaut, W. (2006) 'Conference Report: Epigenetics and Development', *Biological Theory*, 1, 108–9. p. 109.
46. Callebaut, W. and Laubichler, M.D. (2007) 'Biocomplexity as a challenge for biological theory', *Biological Theory*, 2, 1–2.
47. Jablonka, E. and Lamb, M.J. (2005) *Evolution in Four Dimensions*, Harvard, MA: MIT Press, p. 24; see also contributions in Fogel, A., King, B.J. and Shanker, S. (eds) 2008. *Human Development in the Twenty-First Century: Visionary Ideas from Systems Scientists*, Cambridge: Cambridge University Press.

2 Fit for what?

1. Mitchell, M. (2009) *Complexity: A Guided Tour*, Oxford: Oxford University Press. p. 4.
2. Mitchell, M. (2008) 'Five questions', in C. Gershenson (ed.) *Complexity: 5 Questions* Copenhagen: Automatic Press.
3. Mainzer, K. (1997) *Thinking in Complexity: The Complex Dynamics of Matter, Mind and Mankind*, Berlin: Springer.
4. Close, F. (2005) 'Review' of J.D.Barrow, P.C.W. Davies and C.L. Harper Jr. (eds) *Science and Ultimate Reality: Quantum Theory, Cosmology and Complexity*, Cambridge: Cambridge University Press. *Nature*, 434, 438–9.

5. Oyama, S. (1985) *The Ontogeny of Information*, Cambridge: Cambridge University Press.
6. Mauron, A. (2001) 'Is the genome the secular equivalent sof the soul?', *Science*, 291, 831–2.
7. Potts, R. (1997) *Humanity's Descent: The Consequences of Environmental Instability*, New York: Avon Books.
8. Richerson, P.J., Boyd, R.T. and Henrich, J. (2003). 'Cultural evolution of human cooperation', in P. Hammerstein (ed.) *Genetic and Cultural Evolution of Cooperation*, Cambridge, MA: MIT Press.
9. Piaget, J. (1971) *Structuralism*, London: Routledge and Kegan Paul.
10. Ball, P. (2009) *Flow. Nature's Patterns: A Tapestry in Three Parts*, Oxford: Oxford University Press.
11. Prokopenko, M., Boschetti, F. and Ryan, A. (2009) 'An information theoretic primer on complexity, self-organization and emergence', *Complexity*, 15, 11–28.
12. Reading, A. (2007) 'The biological nature of meaningful information', *Biological Theory*, 1, 243–9, p. 247; see also, Victor, J.D. (2006) 'Approaches to information-theoretic analysis of neural activity', *Biological Theory*, 1, 302–16.
13. Stewart, I. (1993) 'Chaos', in Howe, L. and Wain, A. (eds) *Predicting the Future*, Cambridge: Cambridge University Press. p. 30.
14. Mateos, R., Olmedo, E., Sancho, M. and Valderas, J.M. (2002) 'From linearity to complexity: Towards a new economics', *Complexity International*, 10. Paper ID: olmedo02, http://www.complexity.org.au/ci/vol10/olmedo02/, accessed 10 April 2008.
15. Zak, M., Zbilut, J.P. and Meyers, R.E. (1997) *From Instability to Intelligence: Complexity and Predictability in Nonlinear Dynamics (Lecture Notes in Physics: New Series m49)*, New York: Springer Verlag.
16. Mainzer, K. (1997) *Thinking in Complexity: The Complex Dynamics of Matter, Mind and Mankind*, Berlin: Springer, p. 65.
17. Van Orden, G.C., Holden, J.G. and Turvey, M.T. (2003) 'Self-organization of cognitive performance', *Journal of Experimental Psychology: General*, 132, 331–50. p. 344.
18. Prigogine, I. and Stengers, I. (1984) *Order Out of Chaos: New Dialogue with Nature*, New York: Bantam.
19. Taylor, R.P., Micolich, A.P. and Jonas, D. (1999) 'Fractal analysis of Pollock's Drip Paintings', *Nature*, 399, 422.
. Parrott, L., and Kok, R. (2000) 'Incorporating complexity in ecological modeling', *Complexity International*, 7, 1–19.
21. Barrio, R. (2008) 'Turing systems: A general model for complex patterns', in Licata, I. and Sakaji, A. (eds) *Physics of Emergence and Organization*, New York: World Scientific.
22. Prigogine, I., and Nicolis, G. (1998) *Exploring Complexity: An Introduction*, New York: W. H. Freeman.
23. Pokroy, B., Kang, S.H., Mahadevan, L. and Aizenberg, J. (2009) 'Self-organization of a mesoscale bristle into ordered, hierarchical helical assemblies', *Science*, 323, 237–40.
24. Stewart *op cit.* p. 38.
25. Nykter, M., Price, N.D., Aldana, M., Ramsey, S.A., Kauffman, S.A., Leroy E. Hood, L.E., Yli-Harja, O. and Shmulevich, I. (2008) 'Gene expression

dynamics in the macrophage exhibit criticality', *Proceedings of the National Academy of Sciences*, 105, 1897–900.
26. Crutchfield, J.P. (2002) 'What lies between order and chaos', in J. Casti (ed.) *Art and Complexity*, Oxford: Oxford University Press. p. 32.

3 Intelligent life

1. Cleland, C. and Chyba, C. (2002) 'Defining "life"', *Origins of Life and Evolution of the Biosphere*, 32, 387–93. p. 387.
2. Maynard Smith, J. and Szathmary, E. (1999) *The Origins of Life*, Oxford: Oxford University Press.
3. Crick, F. (1981) *Life Itself*, London: Simon and Schuster.
4. Kauffman, S.A. (1996) 'Self-replication: Even peptides do it', *Nature*, 382, 496–7.
5. Segré, D. and Lancet, D. (2000) 'Composing life', *EMBO Reports*, 3, 217–22.
6. Quoted in Ridley, M. (2000) *Mendel's Demon*, Oxford: Oxford University Press.
7. Davies, P. (1998) *The Fifth Miracle: The Search for the Origin of Life*, London: Penguin.
8. de Duve, C. (1995) 'The beginnings of life on earth', *American Scientist*, Sep–Oct, 428–37.
9. Johnson, A.P., Cleaves, H.J., Dworkin, J.P., Glavin, D.P., Lazcano, A. and Bada, J.L. (2008) 'The Miller Volcanic Spark Discharge Experiment', *Science*, 322, 404.
10. Fontecilla-Camps, J.C., Amara, P., Cavazza, C., Nicolet, Y. and Volbeda, A. (2009) 'Structure–function relationships of anaerobic gas-processing metalloenzymes', *Nature*, 460, 814–34.
11. Fox, R.W. (1998) *Energy and the Evolution of Life*, New York: Freeman.
12. Lee, D.H., Granja, J.R., Martinez, J.A., Severin K. and Ghadiri, M.R. (1996) 'Emergence of symbiosis in peptide self-replication through a hypercyclic network', *Nature*, 382, 525–8.
13. Martin, W. and Russell, M. (2002) 'On the origins of cells: A hypothesis for the evolutionary transitions from abiotic geochemistry to chemoautotrophic prokaryotes, and from prokaryotes to nucleated cells', *Philosophical Transactions of the Royal Society, Series B*, 358, 59–85.
14. Shenhav, B., Segre, D. and Lancet, D. (2003) 'Mesobiotic emergence: Molecular and ensemble complexity in early evolution', *Advances in Complex Systems*, 6, 15–35. p. 17.
15. Described in Shenhav et al. *op cit.*
16. Kauffman, S. (1995) *At Home in the Universe, the Search for the Laws of Self-Organization and Complexity*, Oxford: Oxford University Press; see also Kauffman, S.A. (2008) *Reinventing the Sacred*, New York: Edge.
17. Shenhav, S., Solomon, A., Lancet, D. and Kafri, R. (2005) 'Early systems biology and prebiotic networks', in C. Priami (ed.) *Transactions on Computational Systems Biology*, Berlin. Springer-Verlag. p. 14.
18. Stano, P. and Luisi, P.L. (eds) (2007) *International School on Complexity – Basic Questions about the Origins of Life*, Netherlands: Springer; Bich, L. and

Damiano, L. (2007) 'Order in the nothing: Autopoiesis and the organizational characterization of the living', in Licata, I. and Sakaji, A.J. (eds) *Physics of Emergence and Organization*, New York: World Scientific.
19. Shenhav et al. 2003, *op cit*. p. 24.
20. Taylor, W.R. (2005) 'Stirring the primordial soup', *Nature*, 434, 75–6.
21. Eigen, M. and Schuster, P. (1979) *The Hypercycle – A Principle of Natural Self-Organization*, Berlin: Springer-Verlag.
22. Cadenasso, M.L., Pickett, S.T.A. and Grove, J.M. (2006) 'Dimensions of ecosystem complexity: Heterogeneity, connectivity, and history', *Ecological Complexity*, 3, 1–12, 23.
23. Ramanathan, S., and Broach, J. (2007) 'Do cells think?' *Cell. Mol. Life Sci*, 64, 1801–4.
24. Langton, C. (1989) 'Artificial life', in Langton. C. (ed.) *Artificial Life*, Redwood City, CA.: Addison-Wesley. p. 38.
25. Batten, D., Salthe, S. and Boscheti, F. (2008) Visions of evolution: self-organization proposes what natural selection disposes, *Biological Theory*, 3, 17–29.
26. Fox, *op cit*. See also other models discussed by Rasmussen, S., Chen, L., Stadler, B. and Stadler, P. (2004) 'Proto-organism kinetics: Evolutionary dynamics of lipid aggregates with genes and metabolism', *Origins of Life and Evolution of the Biosphere*, 34, 171–81; also Deamer, D., Dworkin, J.P., Sandford, S.A., Bernstein, M.P. and Allamandola, L.J. (2002) 'The first cell membranes', *Astrobiology*, 2, 371–81.
27. Howland, J.L. (2000) *The Surprising Archaea*, Oxford: Oxford University Press.
28. Shenhav, B. and Lancet, D. (2004) 'Prospects of a computational origin of life endeavor', *Origins of Life and Evolution of the Biosphere*, 34, 181–94.
29. Ball, P. (2008) 'Cellular memory hints at the origins of intelligence', *Nature*, 451, 385.
30. Saigusa, T., Tero, A., Nagaki, T and Kuramoto, Y. (2008) 'Amoebae anticipate periodic events', *Physical Review Letters*, 100, 018101.
31. For review see Mello, B.A., Shaw, L. and Tuz, Y. (2004) 'Effects of receptor interaction in bacterial chemotaxis', *Biophysical Journal*, 87, 1578–95.
32. Andrews, S.S. (2005) 'Serial rebinding of ligands to clustered receptors as exemplified by bacterial chemotaxis', *Physical Biology*, 2, 111–22.
33. Wadhams, G.H. and Armitage, J.P. (2004) 'Making sense of it all: Bacterial chemotaxis', *Nature Reviews: Molecular Cell Biology*, 5, 1024–38.
34. Tagkopoulos, I., Liu, Y-C. and Tavazoie, S. (2008) 'Predictive behavior within microbial genetic networks', *Science*, 320, 1313–17. p. 1313.
35. Tagkopoulos, et al., *op cit*, p. 1317.
36. This account is based on Panda, S., Hogenesch, J.B. and Kay, S.A. (2002) 'Circadian rhythms from flies to human', *Nature* 417, 329–5.
37. Gallego, M. and Virshup, D.M. (2007) 'Post-translational modifications regulate the ticking of the circadian clock', *Nature Reviews: Molecular Cell Biology*, 8, 139–48.
38. Voigt, C., Goymann, W. and Leitner, S. (2007) 'Green matters! Growing vegetation stimulates breeding under short-day conditions in wild canaries (*Serinus canaria*)', *Journal of Biological Rhythms*, 22, 554–7.

39. Chen, R., Seo, D.O., Bell, E., von Gall, C. and Lee, C. (2008) 'Strong resetting of the mammalian clock by constant light followed by constant darkness', *Journal of Neuroscience*, 28, 11839–47.
40. De Haro, L. and Satchidananda, P. (2006) 'Systems biology of circadian rhythms: an outlook', *Journal of Biological Rhythms*, 21, 507–18.
41. Delaunay, F. and Laudet, V. (2002) Circadian clocks and microarrays: Mammalian genome gets rhythm, *Trends in Genetics*, 18, 595–6.

4 Bodily intelligence

1. Shankaran H, Resat, H. and Wiley, H.S. (2007) 'Cell surface receptors for signal transduction and ligand transport – a design principles study', *PLoS: Computational Biology*, 3, e101; Cselenyi, C.S. and Lee, E. (2008) 'Context-dependent activation or inhibition of Wnt-β-Catenin signaling', *Science Signaling*, 1, 10.
2. Wiley H.S., Shvartsman, S.Y. and Lauffenburger, D.A. (2003) 'Computational modeling of the EGF-receptor system: A paradigm for systems biology', *Trends in Cell Biology*, 13, 1–43.
3. Hlavacek, W.S. and Faeder, J.R. (2009) 'The complexity of cell signaling and the need for a new mechanic', *Science Signaling*, 2, e46.
4. Ray, L.B. and Gough, N.R. (2002) 'Orienteering strategies for a signaling maze', *Science*, 296, 1632–33.
5. Golub, T., Wacah, S. and Caroni, P. (2003) 'Spatial and temporal control of signaling through lipid rafts', *Current Opinion in Neurobiology*, 14, 542–50.
6. Sachs, K., Perez, O., Pe'er, D., Lauffenburger, D.A. and Nolan, G.P. (2005) 'Causal protein-signaling networks derived from multiparameter single-cell data', *Science*, 308, 523–6.
7. Kashtan, N. and Alon, U. (2005) 'Spontaneous evolution of modularity and network motifs', *Proceedings of the National Acacdemy of Sciences*, 102, 13773–8; Wagner, G.P., Pavlicev, M. and Cheverud, J.M. (2007) 'The road to modularity', *Nature Reviews: Genetics*, 8, 921–6.
8. López-Maury, L. Marguerat, S and Bähler, J. (2008) 'Tuning gene expression to changing environments', *Nature Reviews: Genetics*, 9, 583–94; Pearson, H. (2008) 'The cellular hullabaloo', *Nature*, 453, 150–3. p. 150.
9. Schadt, E.E., Sachs, A. and Friend, S. (2005) 'Embracing complexity, inching closer to reality', *Science STKE*, 295, e40.
10. Vaudry, D., Stork, P.J.S., Lazarovici, P. and Elden, L.E. (2002) 'Signaling pathways for C12 differentiation: Making the right connections', *Science*, 296, 1646–7.
11. Natarajan, M., Lin, K-M., Hsueh, R.C., Sternweis, P.C. and Ranganathan, R. (2006) 'A global analysis of cross-talk in a mammalian cellular signalling network', *Nature Cell Biology*, 8, 571–80. p. 571.
12. Shvartsman, S.Y., Wiley, H.S. and Lauffenburger, D.A. (2004) 'Spatiotemporal dynamics of autocrine loops in the epidermal growth factor receptor system', *IEEE Control Systems Magazine*, August, 53–62.
13. Rosenbaum, D.M., Rasmussen, S.G.F. and Kobilka, B.K. (2009) 'The structure and function of G-protein-coupled receptors', *Nature*, 459, 356–63.

14. Flaherty, P., Radhakrishnan, M.L., Dinh, T., Rebres, R.A., Roach, T.I., et al. (2008) 'A dual receptor crosstalk model of G-protein-coupled signal transduction', *PLoS Computional Biology*, 4, e1000185.
15. Bromley, S.K., Mempel, T.R. and Luster, A.R. (2008) 'Orchestrating the orchestrators: Chemokines in control of T cell traffic', *Nature Immunology*, 9, 970–80.
16. Singleton, K.L., Roybal, K.Y., Sun, Y., Fu, G., Gascoigne, N.R.J., van Oers, N.S.C. and Wülfing, C. (2009) 'Spatiotemporal patterning during T cell activation is highly diverse', *Science Signaling*, 2, ra15.
17. Bestebroer J., van Kessel, K.P.M., Azouagh, H., Walenkamp, A.M., Boer, I.G.J., Romijn, R.A., van Strijp, J.A.G. and de Haas, C.J.C. (2009) 'Staphylococcal SSL5 inhibits leukocyte activation by chemokines and anaphylatoxins', *Blood*, 113, 328–37.
18. Chalfie, M. (2009) 'Neurosensory mechanotransduction', *Nature Reviews: Molecular Cell Biology*, 10, 44–9.
19. Wang, N., Tytell, J.D. and Ingber, D.E. (2009) 'Mechanotransduction at a distance: Mechanically coupling the extracellular matrix with the nucleus', *Nature Reviews: Molecular Cell Biology*, 10, 75–81.
20. Geiger, B., Spatz, J.P. and Bershadsky, A.D. (2009) 'Environmental sensing through focal adhesions', *Nature Reviews: Molecular Cell Biology*, 10, 21–30.
21. Ma, X.M. and Blenis, L. (2009) 'Molecular mechanisms of mTOR-mediated translational control', *Molecular Cell Biology*, 10, 307.
22. Gilbert, W. (1978) 'Why genes in pieces?' *Nature*, 271, 501.
23. Le Hir, H., Nott, A. and Moore, M.J. (2003) 'How introns influence and enhance eukaryotic gene expression', *Trends in Biochemical Sciences*, 28, 215–20.
24. Johnson, J.M., Castle, J., Garrett-Engele, P., Kan, Z., Loerch, P.M., Armour, C.D., Santos, R., Schadt, E.E., Stoughton, R. and Shoemaker, D.D. (2003) 'Genome-wide survey of human alternative pre-mRNA splicing with exon junction microarrays', *Science*, 302, 2141–4.
25. Liu, M. and Grigoriev, A. (2004) 'Genome analysis protein domains correlate strongly with exons in multiple eukaryotic genomes – evidence of exon shuffling?', *Trends in Genetics*, 20, 399–405.
26. Saxonov, S. and Gilbert, W. (2003) 'The universe of exons revisited', *Genetica*, 118, 167–278.
27. Mattick, J.S. (2003) 'Challenging the dogma: The hidden layer of non-protein-coding RNAs in complex organisms', *BioEssays*, 25, 930–939; Storz, G. (2002) 'An expanding universe of noncoding RNAs', *Science* 296, 1260–3; Valadkhan S. and Nilsen T.W. (2010) 'Reprogramming of the noncoding transcriptome during brain development', *Journal of Biology*, 9, 5.
28. Gannon, F. and Pariente, N. (2008) 'Variations on complexity', *EMBO Reports*, 9, 6, 493.
29. Bar-Yam, Y., Harmon, D. and de Bivort, B. (2009) 'Attractors and democratic dynamics', *Science*, 323, 1016–17.
30. Levine, M. and Tjian, R. (2003) 'Transcription regulation and animal diversity', *Nature*, 424, 147–51; Durand, D. (2003) 'Vertebrate evolution: Doubling and shuffling with a full deck', *Trends in Genetics*, 19, 2–5.
31. Nykter, M., Price, N.D., Aldana, M., Ramsey, S.A., Kauffman, S.A., Leroy E. Hood, L.E., Yli-Harja, O. and Shmulevich, I. (2008) 'Gene expression dynamics in the macrophage exhibit criticality', *PNAS*, 105, 1897–900.

32. Wuensche, A. (2002) 'Basins of attraction in network dynamics: A conceptual framework for biomolecular networks', *Santa Fe Instititute Working Paper*, 02-02-004; see also Shankaran, H. and Wiley, H.S. (2008) 'Smad signaling dynamics: Insights from a parsimonious model', *Science Signaling*, 36, e41.
33. Sole, R.V., Cancho, R.F., Montoya, J.M. and Valverde, S. (2002) 'Selection, tinkering and emergence in complex networks', *Santa Fe Working Papers*, 02-07-029.
34. Luscombe, N.M., Babu, M.M., Yu, H., Snyder, M. and Teichmann, S.A. (2004) 'Genomic analysis of regulatory network dynamics reveals large topological changes', *Nature*, 431, 308–12.
35. Wuensche op cit p. 35.
36. Morowitz, H.J., Kosgelnik, J.D., Yang, J. and Coey, G. (2000) 'The origin of intermediary metabolism', *Proceedings of the National Academy of Sciences*, 97, 7704–8.
37. Rose, S. (1997) *Lifelines*, London: Penguin. p. 17.
38. Nederbragt, H. (1997) 'Hierarchical organisation of biological systems and the structure of adaptation in evolution and tumorigenesis', *Journal of Theoretical Biology*, 184, 149–156. p. 151.
39. Glass, L. (1991) 'Nonlinear dynamics of physiological function and control', *Chaos*, 1, 247–50; Shelhamer, M. (2008) *Nonlinear Dynamics in Physiology: A State-Space Approach*, New York: World Scientific.
40. Krain, L.P. and Denver, R.J. (2004) 'Developmental expression and hormonal regulation of glucocorticoid and thyroid hormone receptors during metamorphosis in *Xenopus laevis*', *Journal of Endocrinology*, 181, 91–104. p. 102.
41. Yates, F.E. (2008) 'Homeokinetics/homeodynamics: A physical heuristic for life and complexity', *Ecological Psychology*, 20, 148–79.
42. Joëls, M. and Baram, T.Z. (2009) 'The neuro-symphony of stress', *Nature Reviews: Neuroscience*, 10, 459–66.
43. Shelhamer, M. (2006) *Nonlinear Dynamics in Physiology: A State-Space Approach*, New York: World Scientific.
44. Goldberger, A.L., Amaral, L.A.N., Hausdorff, J.M., Ivanov, P.Ch., Peng, C-K. and Stanley, H.E. (2002) 'Fractal dynamics in physiology: Alterations with disease and aging', *Proceedings of the National Academy of Sciences*, 99 (suppl.1), 2466–72.
45. Costa, M, Goldberger, A.L. and Peng, C-K. (2002) 'Multiscale entropy analysis of complex physiologic time series', *Physical Review Letters*, 89, 068102-1-4.
46. Sabelli, H. and Abouzeid, A. (2002) 'Definition and empirical characterization of creative processes', *Nonlinear Dynamics, Psychology and Life Sciences*, 7, 35–47.
47. Goldberger et al. *op.cit.* p. 2471.

5 Evolution of development

1. Spitzer, N.C. (2009) 'Neuroscience: A bar code for differentiation', *Nature*, 458, 843–4.
2. Pigliucci, M. (2003) 'The new evolutionary synthesis: Around the corner or impossible chimaera?' *The Quarterly Review of Biology*, 78, 449–53; see also,

Lickliter, R. (2008) 'Developmental dynamics: The new view from the life sciences', in Fogel, A., King, B.J. and Shanker, S. (eds) *Human Development in the Twenty-First Century: Visionary Ideas from Systems Scientists*, Cambridge: Cambridge University Press.
3. Müller, G.B. (2008) 'Evo-devo: Extending the evolutionary synthesis', *Nature Reviews: Genetics*, 8, 943–8.
4. Kaiser, D. (2001) 'Building a multicellular organism', *Annual Review of Genetics*, 35, 103–23.
5. Turing, A.M. (1952) 'The chemical basis of morphogenesis', *Philosophical Transactions of the Royal Society, Series B*, 237, 37–72.
6. Wolpert, L. (1969) 'Positional information and the spatial pattern of cellular differentiation', *Journal of Theoretical Biology*, 25, 1–47. See also Holloway, D.M., Reinitz, J., Spirov, A. and Vanario-Alonso, C.E. (2002) 'Sharp borders from fuzzy gradients', *Trends in Genetics*, 18, 385–7; Schier, A.F. and Needleman, D. (2009) 'Developmental biology: Rise of the source–sink model', *Nature*, 461, 480–1.
7. Lewis, J., Hanisch, A. and Holder, M. (2009) 'Notch signaling, the segmentation clock, and the patterning of vertebrate somites', *Journal of Biology*, 8, 44–5.
8. Vogel, G. (2008) 'Breakthrough of the year: Reprogramming cells', *Science*, 322, 1766–7.
9. Weinstein, D.C., and Hemmati-Brivanlou, A. (1999) 'Neural induction', *Annual Review of Cell and Developmental Biology*, 15, 411–33; Lupo, G., Harris, W.A., Barsacchi, G. and Vignali, R. (2002) 'Induction and patterning of the telencephalon in *Xenopus laevis*', *Development*, 129, 5421–36.
10. Chong, L. and Ray, L.B. (2002) 'Whole-istic biology', *Science*, 259, 1661.
11. Schneider, R.A. and Helms, J.A. (2003) 'The cellular and molecular origins of beak morphology', *Science*, 299, 565–8.
12. Nijhout, F. and Emlen, D.J. (1998) 'Competition among body parts in the development and evolution of insect morphology', *Proceedings of the National Academy of Sciences*, 95, 3685–9.
13. Stevens, C.F. (2009) 'Darwin and Huxley revisited: The origin of allometry', *Journal of Biology*, 8(2), 14.
14. Arnold, S.J. and Robertson, E.J. (2009) 'Making a commitment: Cell lineage allocation and axis patterning in the early mouse embryo', *Nature Reviews: Molecular Cell Biology*, 10, 91–103.
15. Coen, E. (1999) *The Art of Genes. How Organisms Make Themselves*, Oxford: Oxford University Press.
16. Patel, N.H. (2004) 'Time, space and genomes', *Nature*, 431, 28–9.
17. Pearson, J.C., Lemons, D. and McGinnis, W. (2005) 'Modulating Hox gene functions during animal body patterning', *Nature Reviews: Genetics*, 6, 893–904.
18. Lall, S. and Patel, N.H. (2001) 'Conservation and divergence in molecular mechanisms of axis formation', *Annual Review of Genetics*, 35, 407–37; Kmita, M. and Duboule, D. (2003) 'Organizing axes in time and space; 25 years of colinear tinkering', *Science*, 301, 333–5.
19. Riechmann, V. and Ephrussi, A. (2001) 'Axes formation during Drosophila oogenesis', *Current Opinion in Genetics and Development*, 11, 374–83.
20. Morisato, D. and Anderson, K.V. (1995) 'Signaling pathways that establish the dorsal-ventral pattern of the *Drosophila* embryo, *Annual Review of Genetics*, 29, 371–99.

21. See for example Lander, A. (2007) 'Morpheus Unbound: Reimagining the morphogen gradient', *Cell*, 128, 245–56; Ochoa-Espinosa, A., Yu, D., Tsirigos, A., Struffi, P. and Small, S. (2009) 'Anterior-posterior positional information in the absence of a strong Bicoid gradient', *Proceedings of the National Academy of Sciences*, 106, 3823–8.
22. Soshnikova, N. and Duboule, D. (2009) 'Epigenetic temporal control of mouse Hox genes *in vivo*', *Science*, 324, 1320–23.
23. Sieweke, M.H. and Graf, T. (1998) 'A transcription factor party in blood cell differentiation', *Current Opinion in Genetics and Development*, 8, 545–51. p. 549.
24. Huang S., Eichler G., Bar-Yam Y. and Ingber D.E. (2005) 'Cell fates as high-dimensional attractor states of a complex gene regulatory network', *Physical Review Letters*, 94, Art. No. 128701.0
25. Flatt, T. (2005) 'The evolutionary genetics of canalization', *The Quarterly Review of Biology*, 80, 287–317.
26. Wilkins, A.S. (2008) 'Canalisation: A molecular genetic perspective', *BioEssays*, 19, 257–62.
27. Schlichting, T.D. and Pigliucci, M. (1998) *Phenotypic Evolution: A Reaction Norm Perspective*. Sunderland, MA: Sinauer.
28. Polaczyk, P.J., Gasperinin, R. and Gibson, G. (1998) 'Naturally occurring genetic variation affects Drosophila photoreceptor determination', *Developmental Genetics and Evolution*, 207, 462–70; Gibson, G. and Wagner, G. (2000) 'Canalization in evolutionary genetics: A stabilizing theory?' *BioEssays*, 22, 372–80.
29. Rutherford, S.L. and Lindquist (1998) 'Hsp90 as a capacitor for morphological evolution', *Nature*, 396, 336–42.
30. Manu and Surkova, S. et al. (2009) 'Canalization of gene expression and domain shifts in the *Drosophila* blastoderm by dynamical attractors', *PLoS Computational Biollogy*, 5, e1000303.
31. Rose, C.R. (2005) 'Integrating ecology and developmental biology to explain the timing of frog metamorphosis', *Trends in Ecology and Evolution*, 20, 129–35.
32. Dent-Read, C. and Zukow-Goldring, P. (1997) 'Epigenetic systems', in C. Dent-Read and P. Zukow-Goldring (eds) *Evolving Explanations of Development*, Washington, D.C.: American Psychological Association. p. 454.
33. Gardner, H. (1984) *Frames of Mind: The Theory of Multiple Intelligences*, London: Heinemann. pp. 56–7.
34. Coen *op cit.*
35. Stearns, S.C. (1989) 'The evolutionary significance of phenotypic plasticity', *BioScience*, 39, 436–47. p. 442.
36. Agrawal A.A., Laforsch, C. and Tollrian, R. (1999) 'Transgenerational induction of defenses in animals and plants', *Nature*, 401, 60–63.
37. Van Buskirk, J. and Relyea, R.A. (1998) 'Selection for phenotypic plasticity in *Rana sylvatica* tadpoles', *Biological Journal of the Linnaean Society*, 65, 301–28.
38. Dodson, S. (1989) 'Predator-induced reaction norms', *BioScience*, 39, 447–52.
39. Piaget, J. (1980) *Adaptation and Intelligence*, Chicago: University of Chicago Press.

40. Nijhout, H.F. (2003) 'Gradients, diffusion and genes in pattern formation', in Müller, G. and Newman, S. (eds) *Origination of Organismal Form*, Cambridge, MA: MIT Press.
41. Gilbert, S.F. (1997) *Developmental Biology. Fifth edition*, Sunderland, MA: Sinauer Associates; Gilbert, S.F. (2001) 'Ecological developmental biology: Developmental biology meets the real world', *Developmental Biology*, 233, 1–12.
42. Crespi, E.J. and Denver, R.J. (2005) 'Ancient origins of human developmental plasticity', *American Journal of Human Biology*, 17, 44–54. p. 51.
43. Horton, T.H. (2005) 'Fetal origins of developmental plasticity: Animal models of induced life history variation', *American Journal of Human Biology*, 17, 34–43; Harper, L.V. (2005) 'Epigenetic inheritance and the intergenerational transfer of experience', *Psychological Bulletin*, 131, 340–60.
44. Richardson, K. and Norgate, S. (2008) 'Behaviour genetic models and realities', *Human Development*, 49, 354–58.
45. Blakemore, C. and Van Sluyters, R.C. (1975) 'Innate and environmental factors in the development of the kitten's visual cortex', *Journal of Physiology*, 248, 663–716.
46. Sur, M. (1993) 'Cortical specification: Microcircuits, perceptual identity, and an overall perspective', *Perspectives on Developmental Neurology*, 1, 109–13.
47. Tropea, D., Van Wart, A. and Sur, M. (2009) 'Molecular mechanisms of experience-dependent plasticity in visual cortex', *Philosophical Transactions of the Royal Society, Series B*, 364, 341–55.
48. Casal, J.J., Fankhauser, C., Coupland, G. and Blazquez, M.A. (2004) 'Signalling for developmental plasticity', *Trends in Plant Science*, 9, 309–15.
49. Crespi and Denver *op.cit.*
50. Rose, C.R. (2005) 'Integrating ecology and developmental biology to explain the timing of frog metamorphosis', *Trends in Ecology and Evolution*, 20, 129–35.
51. Mondor, E.B., Tremblay, M.N. and Lindroth, R.L. (2004) 'Transgenerational phenotypic plasticity under future atmospheric conditions', *Ecology Letters*, 7, 941–946; Mondor, E.B., Tremblay, M.N., Awmack, C.S. and Lindroth, R.L. (2005) 'Altered genotypic and phenotypic frequencies under enriched CO_2 and O_3 atmospheres', *Global Change Biology*, 11, 1990–6.
52. Denver, R.J. (1997) 'Proximater mechanisms of phenotypic plasticity in amphibian metamorphosis', *American Zoologist*, 37, 172–84. p. 174; see also Boorse, G.C. and Denver, R.J. (2004) 'Endocrine mechanisms underlying plasticity in metamorphic timing in spadefoot toads', *Integrative and Comparative Biology*, 43, 646–57.
53. Li, X-Q. (2008) 'Developmental and environmental variation in genomes', *Heredity*, 102, 323–9.
54. Shachar-Dadon, A., Schulkin, J. and Leshem, M. (2009) 'Adversity before conception will affect adult progeny in rats', *Developmental Psychology*, 45, 9–16.
55. Shanks, N., Windle, R.J., Perks, P.A., Harbuz, M.S., Jessop, D.S., Ingram, C.D. and Lightman, S.L. (2000) 'Early-life exposure to endotoxin alters hypothalamic–pituitary–adrenal function and predisposition to inflammation', *Proceedings of the National Academy of Sciences*, 97, 5645–50.
56. Minugh-Purvis, N. and McNamara, K.J. (2002) *Human Evolution through Developmental Change*, New York: Johns Hopkins University Press.

57. Yeh, P.J. and Price, T.D. (2004) 'Adaptive phenotypic plasticity and the successfull colonization of a novel environment', *The American Naturalist*, 164, 531–42; Agrawal, A.A. (2001) 'Phenotypic plasticity in the interactions and evolution of species', *Science*, 294, 321–26; Moczek, A.P. and Nijhout, H.F. (2003) 'Rapid evolution of a polyphenic threshold', *Evolution and Development*, 5, 259–68.
58. Dushek, J. (2002) 'It's the ecology stupid!', *Nature*, 418, 578–9.
59. Baldwin, J.M. (1896) 'A new factor in evolution', *American Naturalist*, 30, 441–451; Turney, P. (1996) 'How to shift bias: Lessons from the Baldwin effect', *Evolutionary Computation*, 4, 271–95.
60. Tollrian, R. and Heibl, C. (2004) 'Phenotypic plasticity in pigmentation in Daphnia induced by UV radiation and fish kairomones', *Functional Ecology*, 18, 497–502.
61. Tramontin, A.D. and Brenowitz, E. A. (2000) 'Seasonal plasticity in the adult brain', *Trends in Neuroscience*, 23, 251–8.
62. Gottlieb, G. (1991) 'Experiential development of behavioral development: theory', *Developmental Psychology*, 27, 4–13. p. 9.
63. Rollo, D.C. (1994) *Phenotypes: Their Epigenetics, Ecology and Evolution*, London: Chapman and Hall.
64. Mayr, E. (1970) *Population, Species and Evolution*, Cambridge, MA: Belknap Press; Bateson, P. (1988) 'The active role of behavior in evolution', in Ho, M-W. and Fox, S.W. (eds) *Evolutionary Processes and Metaphors*, Chichester: Wiley.
65. Mayr, E. (1974) 'Behavior programs and evolutionary strategies', *American Scientist*, 62, 650–9.
66. Purves, D. and Lichtman, W. (1985) *Principles of Neural Development*, Sunderland, MA: Sinauer, p. 141.
67. Goldberg, J.L. (2003) 'How does an axon grow?', *Genes and Development*, 17, 941–58.
68. Levit, P. (2004) 'Sealing cortical cell fate', *Science*, 303, 48–49.
69. Mueller, B.K. (1999) 'Growth cone guidance: First steps towards a deeper understanding', *Annual Review of Neuroscience*, 22, 351–88; Brinks, H., Conrad, S. et al. (2004) 'The repulsive guidance molecule RGMa is involved in the formation of afferent connections in the dentate gyrus', *Journal of Neuroscience*, 24, 3862–9; Huganir, R.L. and Zipursky, S.L. (2004) 'Signaling mechanisms: Editorial overview', *Current Opinion in Neurobiology*, 14, 267–71.
70. Petrovic, M. and Hummel, T. (2008) 'Temporal identity in axonal target layer recognition', *Nature*, 456, 800–3.
71. Grubb, M.S. and Thompson, I.D. (2004) 'The influence of early experience on the development of sensory systems', *Current Opinion in Neurobiology*, 14, 503–512; Weliky, M. and Katz, L.C. (1999) 'Correlational structure of spontaneous neuronal activity in the developing lateral geniculate nucleus in vivo', *Science*, 285, 599–604; Fu, Y-F., Djupsund, K., Gao, H., Hayden, B., Shen, K. and Dan, Y. (2002) 'Temporal specificity in the cortical plasticity of visual space representation', *Science*, 296, 1999–2004.
72. McCormick, D.A. (1999) 'Spontaneous activity: signal or noise?', *Science*, 285, 541–2.
73. Roberts, J.S. (2004) *Embryology, Epigenesis, and Evolution: Taking Development Seriously*, Cambridge: Cambridge University Press.

6 Intelligent eye and brain

1. Masland, R.H. and Martin, P.R. (2007) 'The unsolved mystery of vision', *Current Biology*, 15, R577–83. p. R580.
2. Noë, A. (2002) 'Is the visual world a grand illusion?', *Journal of Consciousness Studies*, 9, 1–12.
3. Kuffler, S.W. (1953) 'Discharge patterns and functional organization of mammalian retina', *Journal of Neurophysiology*, 16, 37–68. For more recent reviews see Treue, S. (2003) 'Climbing the cortical ladder from sensation to perception', *Trends in Cognitive Science*, 7, 469–11; Matthews, G. (2005) 'Making the retina approachable', *Journal of Neurophysiology*, 93, 3034–5.
4. See note 1.
5. Whitney, D., Westwood, D.A. and Goodale, M.A., (2003) 'The influence of visual motion on fast reaching movements to a stationary object', *Nature*, 423, 870–4; Berry, M.J., Brivanlou, M.J. Jordan, T.A. and Meister, M. (1999) 'Anticipation of moving stimuli by the retina', *Nature*, 398, 334–8. See also Whitney, D. (2002) 'The influence of visual motion on perceived position', *Trends in Cognitive Science*, 6, 211–16.
6. Warren, W. H., and Shaw, R. E. (1985). 'Events and encounters as units of analysis for ecological psychology', in W. H. Warren and R. E. Shaw (eds) *Persistence and Change: Proceedings of the First International Conference on Event Perception*, Hillsdale N.J.: Lawrence Erlbaum. p. 6.
7. James, W. (1890) *Principles of Psychology*, New York: Dover.
8. Hoffman, D.D. (1998) '*Visual Intelligence: How We Create What We See*, New York: W.W. Norton.
9. Milner, A.D. and Goodale, M.A. (1995) *Vision in Action*, Oxford: Oxford University Press.
10. Stevenick, R.R., Borst, A. and Bialek, W. (2001) 'Real time encoding of motion: Answerable questions and questionable answers from the fly's visual system', in Zanker, J. M. and Zeil, J. (eds) *Motion Vision – Computational, Neural, and Ecological Constraints*, New York: Springer Verlag.
11. Clifford, C.W.G. and Ibbotson, M.R. (2003) 'Fundamental mechanisms of visual motion detection: Models, cells, and functions', *Progress in Neurobiology*, 68, 409–37.
12. Zihl, J., von Cramon, D., Mai, N. and Schmid, C. (1991) 'Disturbance of movement vision after bilateral posterior brain damage. Further evidence and follow up observations', *Brain*, 114, 2235–52.
13. Wallach, H. and O'Connell, D.N. (1953) 'The kinetic depth effect', *Journal of Experimental Psychology*, 45, 205–17.
14. Richardson, K. and Webster, D.S. (1996) 'Recognition of objects from point light stimuli: Evidence of covariation structures in conceptual representation', *British Journal of Psychology*, 87, 567–91.
15. Coppola, D. and Purves, D. (1996) 'The extraordinarily rapid disappearance of entopic images', *Proceedings of the National Academy of Sciences*, 93, 8001–4.
16. O'Regan, J. K. and Noë, A. (2001) 'A sensorimotor account of vision and visual consciousness', *Behavioral and Brain Sciences*, 24, 939–1011. p. 1011.
17. Barlow, H.R. (1997) 'The knowledge used in vision and where it comes from', *Philosophical Transactions of the Royal Society, Series B*, 35, 1141–7. p. 1141.

18. Lappin, J.S., Tadin, D. and Whittier, E.J. (2002) 'Visual coherence of moving and stationary image changes', *Vision Researc,* 42, 1523–34. p. 1529.
19. Thorpe, S., Fize, D. and Marlot, C. (1996) 'Speed of processing in the human visual system', *Nature,* 381, 520–2.
20. Hegde, J. and Van Essen, D.C. (2000) 'Selectivity for complex shapes in primate visual cortex V2', *Journal of Neuroscience,* 20, RC61; pRC61.
21. Albright, T.D. and Stoner, G.R. (2002) 'Contextual influences on visual processing', *Annual Review of Neuroscience,* 25, 339–79.
22. Kanizsa G. (1979) *'Organization in Vision: Essays on Gestalt Perception',* New York: Praeger; von der Heydt, R., Peterhans, E. and Baumgartner, G. (1984) 'Illusory contours and cortical neuron responses', *Science,* 224, 1260–2.
23. Albright and Stoner *op.cit.* p. 340.
24. Puchalla, J.L., Schneidman, E., Harris, R.A. and Berry, M.J. (2005) 'Redundancy in the population code of the retina', *Neuron,* 46, 493–514.
25. Roska, B. and Werblin, F.S. (2001) 'Vertical Interactions across ten parallel stacked representations in mammalian retina', *Nature,* 410, 583–7; Roska, B., Nemeth, E., Orzo, L. and Werblin, F. (2000) 'Three levels of lateral inhibition: A space-time study of the retina of the tiger salamander', *Journal of Neuroscience,* 20, 1941–51.
26. Olveczky, B.P., Baccus, S.A. and Meister, M. (2003) 'Segregation of object and background motion in the retina', *Nature,* 423, 401–8.
27. Masland, R.H. and Raviola, E. (2000) 'Confronting the complexity: Strategies for understanding the microcircuitry of the retina', *Annual Review of Neuroscience,* 23, 249–84.
28. Dong, D.W. and Atick, J. (1995) 'Statistics of natural time-varying images', *Network Computation in Neural Systems,* 6, 345–58; Dong, D.W. (2001) 'Spatiotemporal inseparability of natural images and visual sensitivities', in Zanker and Zeil, *op.cit.*
29. For review of the way that interactions create motion detection see Hock, H.S., Schöner, G. and Giese, M. (2003) 'The dynamical foundations of motion pattern formation: Stability, selective adaptation, and perceptual continuity', *Perception and Psychophysics,* 65, 429–57.
30. Eckert, M.P. and Zeil, J. (2001) 'Towards an ecology of motion vision', in Zanker and Zeil *op.cit.*
31. Hurri, J. and Hyvarinen, A. (2003) 'Simple-cell-like receptive fields maximize temporal coherence in natural video', *Neural Computation,* 15, 663–91; Hurri, J. and Hyvarinen, A. (2003) 'Temporal and spatial coherence in simple-cell responses: A generative model of natural image sequences', *Network: Computation in Neural Systems,* 1, 527–551. p. 528.
32. Puchalla et al. *op cit.*
33. Brivanlou, I.H., Warland, D.K. and Meister, M. (1998) 'Mechanisms of concerted firing among retinal ganglion cells', *Neuron,* 20, 527–39; Kuhn, A., Aertsen, A. and Rotter, S. (2003) 'Higher-order statistics of input ensembles and the response of simple model neurons', *Neural Computation,* 15, 67–101.
34. Gutig, R., Aharonov, R., Rotter, S. and Sompolinsky, H. (2003) 'Learning input correlations through nonlinear temporally asymmetric Hebbian plasticity', *Journal of Neuroscience,* 23, 3697–714. p. 3707.

35. Renaud, R., Chartier, S. and Albert, G. (2009) 'Embodied and embedded: The dynamics of extracting perceptual visual invariants', in S.J. Guastello, M. Koopmans and D. Pincus (eds) *Chaos and Complexity in Psychology: The Theory of Nonlinear Dynamical Systems*, Cambridge: Cambridge University Press.
36. Godfrey, K.B. and Swindale, N.V. (2007) 'Retinal wave behavior through activity-dependent refractory periods', *PLoS, Computational Biology*, 3, e245.
37. Shamma, S. (2001) 'On the role of space and time in auditory processing', *Trends in Cognitive Sciences*, 5, 340–8. See also Knudson, E.I. (2002) 'Instructed learning in the auditory localization pathway of the barn owl', *Nature*, 417, 322–8; Soto-Faraco, S., Kingstone, A. and Spence, A. (2003) 'Multisensory contributions to the perception of motion', *Neuropsychologia*, 41, 1847–62.
38. Conway, C.M. and Christiansen, M.H. (2005) 'Modality-constrained statistical learning of tactile, visual, and auditory sequences', *Journal of Experimental Psychology: Learning, Memory, and Cognition*, 31, 324–90.
39. Lei, H., Riffell, J.A., Gage, S.L. and Hildebrand, J.G. (2009) 'Contrast enhancement of stimulus intermittency in a primary olfactory network and its behavioral significance', *Journal of Biology*, 8, 21–32.
40. Erisir, A., Horn, S.C.V. and Sherman, S.M. (1997) 'Relative numbers of cortical and brainstem inputs to the lateral geniculate nucleus', *Proceedings of the National Academy of Sciences*, 94, 1517–20.
41. Solomon, S.G., White, A.J.R. and Martin, P.R. (2002) 'Extraclassical receptive field properties of parvocellular, magnocellular, and koniocellular cells in the primate lateral geniculate nucleus', *The Journal of Neuroscience*, 22, 338–49.
42. De Angelis, G.C., Ohzawa, I. and Freeman, R.D. (1995) 'Receptive field dynamics in the central visual pathway', *Trends in Neuroscience*, 18, 451–8.
43. De Angelis et al. *op cit*. p. 452.
44. Lesica, N.A., Ishii, T., Stanley, G.B. and Hosoya, T. (2008) 'Estimating receptive fields from responses to natural stimuli with asymmetric intensity distributions', *PLoS One*, 3, e3060.
45. Butts, D.A., Weng, C., et al. (2007) 'Temporal precision in the neural code and the time scales of natural vision', *Nature*, 449, 92–5.
46. Andolina, I.M., Jones, H.E., Wang, W. and Sillito, A.M. (2007) 'Corticothalamic feedback enhances stimulus response precision in the visual system', *Proceedings of the National Academy of Sciences*, 104, 1685–90.
47. Gilbert, I.D. (1996) 'Plasticity in visual perception and physiology', *Current Opinions in Neurobiology*, 6, 269–74.
48. deCharms, R.C. and Zador, A. (2000) 'Neural representation and the cortical code', *Annual Review of Neuroscience*, 23, 613–47.
49. Samonds, J.M., Zhou, Z., Bernard, M.R. and Bonds, A.B. (2006) 'Synchronous activity in cat visual cortex encodes collinear and cocircular contours', *Journal of Neurophysiology*, 95, 2602–16.
50. For example, Köhler, W. (1930) 'The new psychology and physics', *Yale Review*, 19, 560–76.
51. Mackay, D. (1986) 'Vision: The capture of optical covariation', in Pettigrew, J.D., Sanderson, K.J. and Levick, W.R. (eds) *Visual Neuroscience*, Cambridge: Cambridge University Press.

52. Deco, G. and Lee, T.S. (2004) 'The role of early visual cortex in visual integration: A recurrent neural network model', *European Journal of Neuroscience*, 20, 1089–1100; Lee, T.S., Stepleton, T., Potetz, B. and Samonds, J. (2007) 'Neural encoding of scene statistics for surface and object inference', in S. Dickinson, A. Leonardis, B. Schiele and M. Tarr (eds) *Object Categorization: Computer and Human Vision Perspective*, Cambridge: Cambridge University Press.
53. Buonomano, D.V. and Maass, W. (2009) 'State-dependent computations: spatiotemporal processing in cortical networks', *Nature Reviews: Neuroscience*, 10, 113–25.
54. Di Lollo, V., Enns, J.T. and Rensnik, R.A. (2000) 'Competition for consciousness among visual events: The psychophysics of re-entrant processes', *Journal of Experimental Psychology: General*, 129, 481–507.
55. Blackmore, S.J. (2001) 'Three experiments to test the sensorimotor theory of vision', *Behavioural and Brain Sciences*, 24, 977.
56. Warren, W.H. and Verbrugge, R.R. (1984) 'Auditory perception of breaking and bouncing events: A case study of ecological acoustics', *Journal of Experimental Psychology: Human Perception and Performance*, 10, 704–12.
57. Garcia-Lazaro, J., Ahmed, B. and Schnupp, J. (2006). 'Tuning to natural stimulus dynamics in primary auditory cortex', *Current Biology*, 16, 264–71.
58. Nelken, I. (2004) 'Processing of complex stimuli and natural scenes in the auditory cortex', *Current Opinion in Neurobiology*, 14, 474–480. p. 474.
59. Lewicki, M.S. and Smith, E.C. (2006) 'Efficient auditory coding', *Nature*, 439, 978–82.
60. Gabbiani, F., Krapp, H.G., Koch, C. and Laurent, G. (2002) 'Multiplicative computation in a visual neuron sensitive to looming', *Nature*, 420, 320–4; Schmitt, M. (2001) 'On the complexity of computing and learning with multiplicative neural networks', *Neural Computation*, 14, 241–301; Rothman, J.S., Cathala, L., Steuber, V. and Silver, R.A. (2009) 'Synaptic depression enables neuronal gain control', *Nature*, 457, 1015–18.
61. Morita, K. (2009) 'Computational implications of cooperative plasticity induction at nearby dendritic sites', *Science Signaling*, 2, p. e2.
62. Coba, M.P., Pocklington, A.J., et al. (2009) 'Neurotransmitters drive combinatorial multistate postsynaptic density networks', *Science Signaling*, 2, ra19.
63. Cited by O'Regan, J.K. and Noë, A. (2001) 'A sensorimotor account of vision and visual consciousness', *Behavioral and Brain Sciences*, 24, 1–58.
64. Hilgetag, C.C., and Grant, S. (2000) 'Uniformity, specificity and variability of cortico-cortical connectivity', *Philosophical Transactions of the Royal Socety, Series B*, 355, 7–20.
65. Maguire, E.A., Gadian, D.G., et al. (2000) 'Navigation-related structural change in the hippocampi of taxi drivers', *Proceedings of the National Academy of Sciences*, 97, 4398–403.
66. Elbert, T., Pantev, C., Weinbruch, C., Rockstroh, B. and Taub, E. (1995) 'Increased cortical representation of the fingers of the left hand in string players', *Science*, 270, 305–7.
67. Grubb, M.S. and Thompson, I.D. (2004) 'The influence of early experience on the development of sensory systems', *Current Opinion in Neurobiology*, 14, 503–12.

68. White, L.A., Coppola, D.M. and Fitzpatrick, D. (2004) 'The contribution of sensory experience to the maturation of orientation selectivity in ferret visual cortex', *Nature*, 411, 1049–53.
69. Wang, X. (2004) 'The unexpected consequences of a noisy environment', *Trends in Neurosciences*, 27, 364–6.
70. Clifford, C.W.G. and Ibbotson, M.R. (2003) 'Fundamental mechanisms of visual motion detection: models, cells and functions', *Progress in Neurobiology* 68, 409–37.
71. Freeman, W.F. (2000) 'A proposed name for aperiodic brain activity: Stochastic chaos', *Neural Networks*, 13, 11–13. p. 2.

7 From neurons to cognition

1. Miller, G. (2009) 'On the origin of the nervous system', *Science*, 325, 24–6; Dyerm A.G. and Vuong, Q.C. (2008) 'Insect brains use image interpolation mechanisms to recognise rotated objects', *PLoS ONE*, 3, e4086.
2. Lehrer, M., Srinivasan, M.V., Zhang, S.W. and Horridge, G.A. (1988) 'Motion cues provide the bee's visual world with a third dimension', *Nature*, 356–7. p. 356.
3. Chen, L., Zhang, S. and Srinivasan, M.V. (2003) 'Global perception in small brains: Topological pattern recognition in honeybees', *Proceedings of the National Academy of Sciences*, 100, 6884–9. See also Giurfa, M. (2003) 'Cognitive neuroethology: Dissecting non-elemental learning in a honeybee brain', *Current Opinion in Neurobiology*, 13, 726–35.
4. Lehrer et al. *op cit*; Benard, J., Stach, S. and Giurfa, M. (2006) 'Categorization of visual stimuli in the honeybee *Apis mellifera*', *Animal Cognition*, 9, 257–70.
5. Greenspan, R.J. and van Swinderen, B. (2004) 'Cognitive consonance: Complex brain functions in the fruit fly and its relatives', *Trends in Neurosciences*, 27, 707–11.
6. Richardson, K. (1999) *Models of Cognitive Development*, Hove: Psychology Press.
7. Turing, A.M. (1950) Computing machinery and intelligence, *Mind*, 59, 433–60.
8. Fodor, J. (2000) *The Mind Doesn't Work That Way: The Scope and Limits of Computational Psychology*, Cambridge, MA: MIT Press.
9. McClelland, J.L., Rumelhart, D.E. and the PDP Research Group (1986) *Parallel Distributed Processing*, Cambridge, Mass.: MIT Press.
10. See contributions in Spencer, J., Thomas, M.S.C. and McClelland, J.L. (eds) (2009) *Toward a New Grand Theory of Development: Connectionism and Dynamical Systems Theory Re-Considered*, Oxford: Oxford University Press.
11. Neisser, U. (2006) 'Foreword', in Spivey, M.J., *Continuity of Mind*, Oxford: Oxford University Press.
12. Spivey, M.J. and Dale, R. (2006) 'Continuous dynamics in real-time cognition', *Current Directions in Psychological Science*, 15, 207–11.
13. Freeman, W.J. (2009) 'The neurobiological infrastructure of natural computing: Intentionality', *New Mathematics and Natural Computation*, 5, 19–30. p. 19.

14. Skarda, C. and Freeman, W.J. (1987) 'How brains make chaos in order to make sense of the world', *Behavioral and Brain Sciences*, 10, 161–95.
15. Freeman, W.J. (2001) *Neurodynamics. An Exploration of Mesoscopic Brain Dynamics*, London: Springer.
16. Beggs, J.M. (2008) 'The criticality hypothesis: How local cortical networks might optimize information processing', *Philosophical Transactions of the Royal Society, Series A*, 366, 329–43; Hsu, D. and Beggs, J.M. (2006) 'Neuronal avalanches and criticality: A dynamical model for homeostasis', *Neurocomputing*, 69, 1134–6.
17. Holden, J.G. (2008) 'Gauging the fractal dimension of cognitive performance', in Riley, M.A. and Van Orden, G.C. (eds) *Tutorials in Contemporary Nonlinear Methods for the Behavioral Sciences*, Arlington: National Science Foundation.
18. Stephen, D.G., Boncoddoa, R., Magnuson, J.S. and Dixon, J.A. (2009) 'The dynamics of insight: Mathematical discovery as a phase transition', *Memory and Cognition*, 37, 1132–49. p. 2.
19. Siri, B., Quoy, M., Delord, B., Cessac, B. and Berry, H. (2007) 'Effects of Hebbian learning on the dynamics and structure of random networks with inhibitory and excitatory neurons', *Journal of Physiology (Paris)*, 101,136–48. For review see, Werner, G. (2007) 'Metastability, criticality and phase transitions in brain and its models', *Biosystems*, 90, 496–508.
20. Molter, C., Salihoglu, U. and Bersini, H. (2006) 'How to prevent spurious data in a chaotic brain', *Proceedings of the IJCNN Conference*, Vancouver: IEE Press: Proc. WCCI,1365–71.
21. Freeman, W.J. (2000) 'Brains create macroscopic order from microscopic disorder by neurodynamics in perception', in P. Århem, C. Blomberg and H. Liljenström, (eds) *Disorder versus Order in Brain Function: Essays in Theoretical Neurobiology*, Singapore: World Scientific Publishing Co.
22. Freeman, W.J. (1995) *Societies of Brains*, Hillsdale, NJ: Erlbaum. p. 67. See also Freeman, W.J. (1999) *How Brains Make up Their Minds*, London: Weidenfeld and Nicolson.
23. Wong, K. and Huk, A.C. (2008) 'Temporal dynamics underlying perceptual decision making: Insights from the interplay between an attractor model and parietal neurophysiology', *Frontiers in Neuroscience*, 2, 245–56.
24. Richardson, K. and Webster, D.S. (1996) 'Recognition of objects from pointlight stimuli: Evidence for covariation hierarchies in conceptual representation', *British Journal of Psychology*, 87, 567–91.
25. Richardson and Webster *op.cit*.
26. Richardson, K. (1992) 'Covariation analysis of knowledge representation: Some developmental studies', *Journal of Experimental Child Psychology*, 53, 129–50.
27. Gregory, R. (2004) 'The blind leading the sighted: An eye-opening experience of the wonders of perception', *Nature*, 430, 836.
28. Peelen, M.V., Fei, L.F. and Kastner, S. (2009) 'Neural mechanisms of rapid natural scene categorization in human visual cortex', *Nature*, 460, 94–7. p. 94.
29. Peelen et al. *op cit* p. 94.
30. Velichkovsky, B.M. (2007) *Towards an Evolutionary Framework for Human Cognitive Neuroscience*, Cambridge, MA: MIT Press.

31. Calvert, G.A., Spence, C. and Stein, B.E. (eds) (2004) *The Handbook of Multisensory Processes*, New York: Bradford Books.
32. Conway, M.C. and Christiansen, M.H. (2005) 'Modality-constrained statistical learning of tactile, visual, and auditory sequences', *Journal of Experimental Psychology: Learning, Memory, and Cognition*, 31, 24–39.
33. Nelken, I. (2004) 'Processing of complex stimuli and natural scenes in the auditory cortex', *Current Opinion in Neurobiology*, 14, 474–80.
34. Blake, R., Sobel, K.V. and James, T.W. (2004) 'Neural synergy between kinetic vision and touch', *Psychological Science*, 15, 397–403.
35. Teramoto, W., Hidaka, S., Gyoba, J. and Suzuki, Y. (2008) 'Dynamic auditory cues modulate visual motion processing', *Perception*, 37 (ECVP Abstract Supplement), 72.
36. McGurk, H. and MacDonald, J. (1976) 'Hearing lips and seeing voices', *Nature*, 264, 746–8.
37. Soto-Faraco, S., Kingstone, A. and Spence, C. (2003) 'Multisensory contributions to the perception of motion', *Neuropsychologia*, 41, 1847–62.
38. Jackson, S.R. (2001) 'Action binding: Dynamic interactions between vision and touch', *Trends in Cognitive Sciences*, 5, 505–6.
39. Wallace, M.T., Meredith, M.A. and Stein, B.E. (1992) 'Integration of multiple sensory modalities in cat cortex', *Experimental Brain Research*, 91, 484–8; see also Miyashita, Y. (2004) 'Cognitive memory: Cellular and network machineries and their top-down control', *Science*, 306, 14–16.
40. Calvert, G.A. and Campbell, R. (2003) 'Reading speech from still and moving faces', *Journal of Cognitive Neuroscience*, 15, 57–70; Munhall, K.G. and Buchan, J.N. (2004) 'Something in the way she moves', *Trends in Cognitive Sciences*, 8, 51–3.
41. Ernst, M.O. (2005) 'The "puzzle" of sensory perception: Putting together multisensory information', *Proceedings of the 7th International Conference on Multimodal Interfaces*, New York: ACM Press.
42. Hubbard, T.L. (1995) 'Environmental invariants in the representation of motion: Implied dynamics and representational momentum, gravity, friction and centripedal force', *Psychonomic Bulletin and Review*, 2, 322–38.
43. Kerzel, D. (2002) 'A matter of design: No representational motion without predictability', *Visual Cognition*, 9, 66–80. p. 78.
44. Nelson, K. (1986) *Event Knowledge*, Hilldale, NJ: Erlbaum.
45. Zacks, J.M. and Tversky, B. (2001) 'Event structure in perception and conception', *Psychological Bulletin*, 127, 3–22.
46. Speer, N.K., Swallow, K.M. and Zachs, J.M. (2003) 'Activation of human motion processing areas during event perception', *Cognitive, Affective, and Behavioral Neuroscience*, 3, 333–45.
47. Chen, J.L., Penhune, V.B. and Zatorre, R.J. (2008) 'Listening to musical rhythms recruits motor regions of the brain', *Cerebral Cortex*, 18, 2844–54.
48. Harman, K.L., Humphrey, G.K. and Goodale, M.A. (2003) 'Active manual control of object views facilitates visual recognition', *Current Biology*, 9, 1315–19.
49. Jackson, S.R. (2001) 'Action binding: Dynamic interactions between vision and touch', *Trends in Cognitive Science*, 5, 505–6.
50. Jackson 2001 op cit. p. 505.
51. For brief review see Di Pellegrino, G. (2001) 'Vision and touch in parietal area', *Trends in Cognitive Science* 5, 48–50.

52. Gallese, V. (2000) 'The inner sense of action: Agency and motor representations', *Journal of Consciousness Studies*, 7, 23–40. See also Kawato, M. (1999) 'Internal models for motor control and trajectory planning', *NeuroReport*, 9, 718–27.
53. O'Regan, J.K. and Noë, A. (2001) 'A sensorimotor account of vision and visual consciousness', *Behavioral and Brain Sciences*, 24, 1–58.

8 Cognitive functions

1. See review in Pylyshyn Z.W. (2002) 'Mental imagery: In search of a theory', *Behavioral and Brain Sciences*, 25, 157–237.
2. Fodor, J. (1994) 'Concepts: A potboiler', *Cognition*, 50, 95–113. p. 95.
3. Harnad, Stevan (2005) 'To cognize is to categorize: Cognition is categorisation', in Lefebvre, C. and Cohen, H. (eds) *Handbook of Categorization in Cognitive Science*, New York: Elsevier.
4. Giurfa, M. (2009) *Animal Cognition: Nonelemental Learning Beyond Simple Conditioning*, CSH Monographs, Cold Spring Harbor: Cold Spring Harbor Laboratory Press. p. 281.
5. Hebb, D.O. (1949) *The Organization of Behavior*, New York: Wiley. p. 14.
6. Zanone, P-G. and Kostrubiec, V. (2004) 'Searching for dynamic principles of learning', in V.K. Jirsa and J.A.S. Kelso (eds) *Coordination Dynamics, Issues and Trends*, New York: Springer.
7. For example, Becerikli, Y., Konar, A.F. and Samad, T. (2003) 'Intelligent optimal control with dynamic neural networks', *Neural Networks*, 16, 251–9.
8. Shiffrin, R.M. (2003) 'Modelling perception and memory', *Cognitive Science*, 27, 341–78.
9. Nader, K. (2003) 'Memory traces unbound', *Trends in Neurosciences*, 26, 65–75. p. 70. See also Tulving, E. (2002) 'Episodic memory: From mind to brain', *Annual Review of Psychology*, 51, 1–25.
10. Shiffrin op. cit. p. 375.
11. Eysenck, M. (2004) *Psychology: An International Perspective*, London: Psychology Press.
12. Elman, J., Bates, E., Karmiloff-Smith, A., Johnson, M., Parisi, D. and Plunkett, K. (1997) *Rethinking Innateness: A Connectionist Perspective on Development*, Cambridge, Mass: MIT Press. p. 359.
13. Chomsky, N. (1980) *Rules and Representations*, Oxford: Blackwell.
14. Pinker, S. (1997) *How the Mind Works*, London: Penguin. p. 21.
15. Pothos, E.M. (2003) 'The rules versus similarity distinction', *Behavioral and Brain Sciences*, 28, 1–49. p. 26.
16. Barsch, R. (2002) *Consciousness Emerging*, Amsterdam: John Benjamins. p. 38.
17. Eysenck, M.W. and Keane, M.T. (1990) *Cognitive Psychology*, Hove: Erlbaum. p. 462.
18. Shepard, R.N. and Metzler, J. (1976) 'Mental rotation of three-dimensional objects', *Science*, 171, 701–703.
19. Pinker op.cit. p. 88.
20. MacLennan, B.J. (2004) 'Natural computation and non-Turing models of computatio', *Theoretical Computer Science*, 317, 115–45.
21. Fodor, J. (2001) *The Mind Doesn't Work That Way: The Scope and Limits of Computational Psychology* Cambridge, Mass.: MIT Press.

22. Eysenck, M.W. (2004) *Psychology: An International Perspective*, Hove: Psychology Press.
23. Richardson, K. (2002) 'What IQ tests test', *Theory and Psychology*, 12, 283–314.
24. Eysenck op.cit. p. 371.
25. See contributuions in P.A. Frensch and J. Funke (eds) (2005), *Complex problem solving: The European Perspective*, Hillsdale, NJ: Lawrence Erlbaum Associates.
26. for demonstration in ANNs see Ciszac, M., Montina, A. and Arecchi, F.T. (2009) 'Control of transient synchronization with external stimuli', *Chaos*, 19, 015104; doi:10.1063/1.3080195.
27. Manrubia, S.C., Mikhailov, A.S. and Zanette, D.H. (2004) *Emergence of Dynamical Order: Synchronization Phenomena in Complex Systems*, Singapore: World Scientific; Kaneko, K. and Tsuda, I. (2001) *Complex Systems: Chaos and Beyond*, Berlin: Springer.
28. Levine, D.S. (2009) 'How does the brain create, change, and selectively override its rules of conduct?', in Perlovky, L.I. and Kozma, R. (eds) *Neurodynamics of Cognition and Comsciousness*, New York: Springer.
29. Phelps, E.A. and LeDoux, J.E. (2005) 'Contributions of the amygdala to emotion processing: From animal models to human behavior', *Neuron*, 48, 175–187.
30. Vuilleumier, P. and Huang, Y-M. (2009) 'Emotional attention: Uncovering the mechanisms of affective biases in perception', *Current Directions in Psychological Science*, 18, 148–52.
31. Desmurget, M., Epstein, C.M., et al. (1999) 'Role of the posterior parietal cortex in updating reaching movements to a visual target', *Nature Neuroscience*, 2, 563.
32. Pessoa, L. (2008) 'On the relation between emption and cognition', *Nature Reviews: Neuroscience*, 9, 148–55; Damaraju E, Huang Y.M., Barrett L.F. and Pessoa, L. (2009) 'Affective learning enhances activity and functional connectivity in early visual cortex', *Neuropsychologia*, 47, 2480–7.
33. Roesch, M.R. and Olson, C.R. (2004) 'Neuronal activity related to reward value and motivation in primate frontal cortex', *Science*, 304, 15–18.
34. Perlovsky and Kozma op cit p. 5.
35. Guastello, S.J. (2006) 'Motor control research requires nonlinear dynamics', *American Psychologist*, 61, 77–8.
36. See note 32.
37. Piaget, J. (1988) 'Piaget's theory', in K. Richardson and S. Sheldon (eds) *Cognitive Development to Adolescence*, Hove: Erlbaum. p. 11.
38. Piaget, J. (1971) *Structuralism*, London: Routledge and Kegan Paul.
39. Doré, F.Y. and Dumas, C. (1987) 'Psychology of animal cognition: Piagetian studies', *Psychological Bulletin*, 102, 219–33.
40. Wynne, C. (2006) *Do Animals Think?* Princeton: Princeton University Press.
41. Tsuda, I. (2001) 'Towards an interpretation of dynamic neural activity in terms of chaotic dynamical systems', *Behavioral and Brain Sciences*, 24, 793–810; Kaneko, K. and Tsuda, I. (2003) 'Chaotic itinerancy', *Chaos*, 13, 926–36.

42. Siri, B., Quoy, M. et al. (2007) 'Effects of Hebbian learning on the dynamics and structure of random networks with inhibitory and excitatory neurons', *Journal of Physiology-Paris*, 10, 136–48.
43. Kaneko, K. (1990) 'Clustering, coding, hierarchical ordering and control in a network of chaotic elements', *Physica*, 41(D), 137–72; see also Anderson, J.A. and Sutton, J.P. (1995) 'A network of networks: Computation and neurobiology', *World Congress on Neural Networks*, 1, 561–8; Sutton, J.P. (1997) 'Network hierarchies in neural organization, development and pathology', in Lumsden, C.L., Brandts, W. A. and Trainor, L.E.H. (eds) *Physical Theory in Biology*, Singapore: World Scientific.
44. Friston, K. and Kiebel, S. (2009) 'Cortical circuits for perceptual inference', Neural Networks, 22, 1093–1104; Friston, K. and Kiebel, S. (2009) 'Attractors in song', *New Mathematics and Natural Computation*, 5, 83–114; Bressler, S.L. and Kelso, J.A.S. (2001) 'Cortical coordination dynamics and cognition', *Trends in Cognitive Science*, 5, 26–36.
45. Tuller, B. (2007) 'Categorization and learning in speech perception as dynamical processe', in Riley, M.A and Van Orden, G. (eds) *Tutorials in Contemporary Nonlinear Methods for the Bahavioral Sciences*, Washington D.C.: National Science Foundation. p. 392.
46. Thelen, E. and Smith, L.B. (1994) *A Dynamic Systems Approach to the Development of Cognition and Action*, Cambridge, MA: MIT Press.
47. see, for example, Kotrschal K, van Staaden, M.J. and Huber, R. (1998) 'Fish brains: evolution and functional relationships', *Review of Fish Biology and Fisheries*, 8, 373–408; Watson, R. (2009) Selectivity for conspecific vocalizations within the primate insular cortex, *Journal of Neuroscience*, 29, 6769–70.
48. Clark, A. and Chalmers, D. (1998) 'The extended mind', *Analysis*, 58, 7–19; Menary, R. (2007) *Cognitive Integration: Mind and Cognition Unbounded*, London: Palgrave Macmillan.
49. Spivey, M., Richardson, D. and Fitneva, S. (2004). 'Thinking outside the brain: Spatial indices to linguistic and visual information', in Henderson, J. and Ferreira, F. (eds), *The Interface of Vision Language and Action*, New York: Psychology Press. pp. 161–2.
50. Pincus, D. (2009) 'Fractal thoughts on fractal brains', SCPTLS Blog, 4/9/09.
51. Stein, L.A. (1999) 'Challenging the computational metaphor: Implications for how we think', *Cybnernetics and Systems*, 30: 4–30. pp. 6–7.
52. Freeman, W.J. (1999) 'Noise-induced first-order phase transitions in chaotic brain activity', *International Journal of Bifurcation and Chaos*, 9, 2215–8. p. 2218.
53. Fitch, R.H., Miller, S. and Tallal, P. (1997). Neurobiology of speech perception, *Annual Review of Neuroscience*, 20, 33153.
54. Chartier, S., Renaud, P. and Boukadoum, M. (2008) 'A nonlinear dynamic artificial neural network model of memory', *New Ideas in Psychology*, 26, 252–77; Nader *op cit*.
55. Bartlett, F.C. (1958) *Thinking: An Experimental and Social Study*, London, Allen and Unwin. p. 200.
56. Velichkovsky, B.M. (2007) *Towards an Evolutionary Framework for Human Cognitive Neuroscience*, Cambridge, MA: MIT Press. p. 3.

57. Sabelli, H and Abouzeid, A. (2003) 'Definition and empirical characterization of creative processes', *Nonlinear Dynamics, Psychology, and Life Sciences*, 7, 35–48.

9 Social intelligence

1. See contributions in Dworkin, M. and Kaiser, D. (eds) (2000) *Myxobacteria II*, Washington DC: American Society for Microbiology.
2. Queller, D.C., Ponte, E., Bozzaro, S. and Strassmann, J.E. (2003) 'Single-gene greenbeard effects in the social amoeba, *Dictyostelium discoideum*', *Science*, 299, 105–6.
3. Velicer, G.J. and Stredwick, K.L. (2002) 'Experimental social evolution with *Myxococcus xanthus*', *Antonie van Leeuwenhoek*, 81, 155–64.
4. Ostrowski E.A., Katoh. M., Shaulsky, G., Queller, D.C. and Strassmann, J.E. (2008) 'Kin discrimination increases with genetic distance in a social amoeba', *PLoS Biology*, 6, e287.
5. Scharf, M.E., Wu-Scharf, D., Pittendrigh, B.R. and Bennett, G.W. (2003) 'Caste- and development-associated gene expression in a lower termite', *Genome Biology*, 4, 1–11.
6. Theraulaz, G., Banabeau, E., et al. (2002) 'Spatial patterns in ant colonies', *Proceedings of the National Academy of Sciences*, 99, 9645–9.
7. Theraulaz et al., *op.cit.* p. 9648.
8. Sulis, W. (2009) 'Collective intelligence: observations and models', in Guastello, S.J., Koopmans, M. and Pincus, D. (eds) *Chaos and Complexity in Psychology*, Cambridge: Cambridge University Press. p. 47.
9. Szuba, T. (2001) *Computational Collective Intelligence*, London: Wiley.
10. Miramontes, O. (1995) 'Order-disorder transitions in the behavior of ant societies', *Complexity*, 1, 56–60.
11. Theraulaz, G. Bonabeau, E., et al. *op cit*.
12. Vandermeer, J., Perfecto, I. and Philpott, S.M. (2008) 'Clusters of ant colonies and robust criticality in a tropical agroecosystem', *Nature*, 451, 457–9.
13. Guerin, S. and Kunkle, D. (2004) 'Emergence of constraint in self-organizing systems', *Nonlinear Dynamics, Psychology, and Life Sciences*, 8, 131–147. p. 133.
14. Brown, C. and Laland, K.N. (2003) 'Social learning in fishes: A review', in Brown, C., Laland, K.N. and Krause, J. (eds) *Learning in Fishes: Why They Are Smarter Than You Think. Fish and Fisheries*, 4, 280–88. p. 280.
15. Grünbaum, D., Viscido, S. and Parrish, J.K. (2005) 'Extracting interactive control algorithms from group dynamics of schooling fish', in Kumar, V., Leonard, N.E. and Morse, A.S. (eds) *Cooperative Control (Proceedings of the Block Island Workshop)*, New York: Springer-Verlag; Tien, J.H., Levin, S.A. and Rubenstein, D.I. (2004) 'Dynamics of fish shoals: Identifying key decision rules', *Evolutionary Ecology Research*, 6, 555–65.
16. Chase, I.D., Tovey, C., Spangler-Martin, D. and Manfredonia, M. (2002) 'Individual differences versus social dynamics in the formation of animal dominance hierarchies', *Proceedings of the National Academy of Sciences*, 99, 5744–9.

17. Hewitt, S.E., Macdonald, D.W. and Dugdale, H.L. (2009) 'Context-dependent linear dominance hierarchies in social groups of European badgers *Meles meles*', *Animal Behaviour*, 77, 161–9.
18. Adolphs, R. (2001) 'The neurobiology of social cognition', *Current Opinion in Neurobiology*, 11, 231–9. p. 231.
19. Humphreys, G.W. and Forde, E.M.E. (2000) 'Hierarchies, similarity and interactivity in object recognition: On the multiplicity of "category-specific" deficits in neuropsychological populations', *Behavioral and Brain Sciences*, 24, 453–509.
20. Watve, M. (2002) 'Bee-eaters (*Merops orientalis*) respond to what a predator can see', *Animal Cognition*, 5, 253–9.
21. Emery, N.J. and Clayton, N.S. (2001) 'Effects of experience and social context on prospective caching strategies by scrub jays', *Nature*, 414, 443–6.
22. Paz-y-Miño, C.G., Bond, A.B., Kamil, A.C. and Balda, R.P. (2004) 'Pinyon jays use transitive inference to predict social dominance', *Nature*, 430, 778–88.
23. Shettleworth, S.J. (2009) *Cognition, Evolution, and Behavior*, Oxford: Oxford University Press.
24. Kuroshima, H., Fujita, K., Fuyuki, A. and Masuda, T. (2002) 'Understanding of the relationship between seeing and knowing by tufted capuchin monkeys (*Cebus apella*)', *Animal Cognition*, 5, 41–8. See also Zimmer, C. (2003) 'How the mind reads other minds', *Science*, 300, 1079–80.
25. For review see Dunbar, R.I.M. (2003) 'The social brain: Mind, language, and society in evolutionary perspective', *Annual Review of Anthropology*, 32, 163–81.
26. Humphrey, N.K. (1976) 'The social function of intellect', in Bateson, P.P.G. and Hinde, R.A. (eds) *Growing Points in Ethology*, Cambridge: Cambridge University Press.
27. See contributions in Byrne, R.W. and Whiten, A. (eds) (1988) *Machiavellian Intelligence*, Oxford: Clarendon Press.
28. Bjorklund, D.F., Yunger, J.L., Bering, J.M. and Ragan, P. (2002) 'The generalization of deferred imitation in enculturated chimpanzees (*Pan troglodytes*)', *Animal Cognition*, 5, 49–58.
29. Whiten, A., Horner, V. and de Waal, F.B.M. (2005) 'Conformity to cultural norms of tool use in chimpanzees', *Nature*, 437, 737–40.
30. Zentall, T.R. (2003) 'Imitation by animals: How do they do it?', *Current Directions in Psychological Science*, 12, 91–6.
31. Boysen, S.T and Himes, G.T. (1999) 'Current issues and emergent theories in animal cognition', *Annual Reviews in Psychology*, 50, 683–705. p. 687.
32. Byrne, R.W., Barnard, P.J., Davidson, I., Janik, V.M., McGrew, W.C., Miklósi, Á. and Wiessner, P. (2004) 'Understanding culture across species', *Trends in Cognitive Sciences*, 8, 341–6.
33. Bjorklund et al. *op.cit.* p. 56.
34. Rowell, T. (2005) 'The myth of peculiar primates', in Box, H.O. and Gibson, K.R. (eds) *Mammalian Social Learning: Comparative and Ecological Perspectives*, Cambridge: Cambridge University Press.
35. Gallese, V. (2009) 'Mirror Neurons', in Baynes, T., Cleeremans, A. and Wilken, P. (eds) *The Oxford Companion to Consciousness*, Oxford: Oxford University Press.

36. Gallese, V., Keysers, C. and Rizzolatti, G. (2004) 'A unifying view of the basis of social cognition', *Trends in Cognitive Sciences*, 8, 396–404.
37. Bush, E.C. and Allman, J.M. (2004) 'The scaling of frontal cortex in primates and carnivores', *Proceedings of the National Academy of Sciences*, 101, 3962–6. See also Reader, S.M. and Laland, K.N. (1999) 'Forebrain size, opportunism and the evolution of social learning in nonhuman primates', *Ethology*, 34 (Supplement), 50.
38. McKinney, M.L. and McNamara, K.J. (1990) *Heterochrony: The Evolution of Ontogeny*, New York: Plenum Press.
39. Dunbar, R. (1998) 'The social brain hypothesis', *Evolution and Anthropology*, 6, 178–90; Dunbar, R.I.M. (2003) 'The social brain: mind, language, and society in evolutionary perspective', *Annual Review of Anthropology*, 32, 163–81.
40. Finarelli, J.A. and Flynn, J.J. (2009) 'Brain-size evolution and sociality in Carnivora', *Proceedings of the National Academy of Sciences*, 106, 9345–9.
41. Dunbar (2003) op cit, p. 168.
42. Reader, S.M. and Laland, K.N. (2002) 'Social intelligence, innovation and enhanced brain size in primates', *Proceedinsg of the National Academy of Sciences*, 99, 4436–41.
43. Jerison, H. (1993) 'The evolved mind', *Behavioral and Brain Sciences*, 16, 763–4. See also contributions in Falk, D. and Gibson, K.R. (eds) (2006) *Evolutionary Anatomy of the Primate Cerebral Cortex*, Cambridge: Cambridge University Press.
44. Allman, J.M., Hakeem, A., Erwin, J.M., Nimchinsky, E. and Hof, P. (2001) 'Anterior cingulated cortex: the evolution of an interface between emotion and cognition', *Annals of the New York Academy of Sciences*, 935, 107–17.
45. Premack, D. and Premack, A.J. (1996) 'Why animals lack pedagogy and some cultures have more of it than others, in Olson, D.R. and Torrance, N. (eds) *The Handbook of Human Development and Education*, Oxford: Blackwell.
46. Tomasello, M. and Warnekan, F. (2008) 'Human behaviour: Share and share alike', *Nature*, 454, 1057–8.
47. Carruthers, P. (2002) 'The cognitive functions of language', *Behavioral and Brain Sciences*, 25, 657–726.
48. Bshary, R., Wickler, W. and Fricke, H. (2002) 'Fish cognition: a primate's eye view', *Animal Cognition*, 5, 1–13. p. 5. See also Dugatkin, L.A. (1997) *Cooperation Among Animals*, Oxford: Oxford University Press.
49. Finarelli and Flynn *op cit*.

10 Intelligent humans

1. Byrne, R.W., Barnard, P.J., et al. (2004) 'Understanding culture across species', *Trends in Cognitive Sciences*, 8, 341–6. p. 341.
2. Herculano-Houzel, S. (2009) 'The human brain in numbers: a linearly scaled-up primate brain', *Frontiers in Human Neuroscience*, 3, 31; Roth, G. and Dicke, U. (2005) 'Evolution of the brain and intelligence', *Trends in Cognitive Sciences*, 9, 250–7.
3. Premack, D. and Premack, A. J. (1996) 'Why animals lack pedagogy and some cultures have more of it than others, in Olson, D.R. and Torrance, N. (eds) *The Handbook of Human Development and Education*, Oxford: Blackwell.

4. Tomasello, M. and Warnekan, F. (2008) 'Human behaviour: Share and share alike', *Nature*, 454, 1057–8.
5. Zollikofer, C.B.E., de Leon, M.S.P., et al. (2005) 'Virtual cranial reconstruction of *Sahelanthropus tchadensis*', *Nature*, 434, 755–9.
6. Wood, B. (2002) 'Palaeoanthropology: Hominid revelations from Chad', *Nature*, 418, 133–5.
7. See note 5.
8. White, T.D., Asfaw, B., et al. (2009) '*Ardipithecus ramidus* and the paleobiology of early hominids', *Science*, 326, 64–78.
9. Rakic, P. (2009) 'Evolution of the neocortex: A perspective from developmental biology', Nature Reviews: Neuroscience, 10, 724–44.
10. Mellars, P. (2004) 'Neanderthals and the modern human colonization of Europe', *Nature*, 432, 461–6.
11. Kaessmann, H., Weibe, V. and Paabo, S. (1999), 'Extensive nuclear DNA sequencing diversity among chimpanzees', *Science*, 286, 1159–61.
12. Manica, A., Amos, W., Balloux, F. and Hanihara, T. (2007) 'The effect of ancient population bottlenecks on human phenotypic variation', *Nature*, 448, 346–8.
13. See contributions in Fuchs, A. and Jirsa, V.K. (2008) *Coordination: Neural, Behavioral and Social Dynamics*, New York: Springer.
14. Feldman, J. (1993) 'From mimesis to synthesis', *Behavioral and Brain Sciences*, 16, 759.
15. Zukow-Goldring, P. (1997) 'A social ecological realist approach to the emergence of the lexicon', in Dent-Read, C. and Zukow-Goldring, P. (eds) *Evolving Explanations of Development*. Washington, DC: American Psychological Association. p. 210.
16. Abraham, F.D. (1992) 'Chaos, bifurcations, and self-organization', *Psychoscience*, 1, 85–118.
17. Sawyer, R.K. (2004) 'The mechanisms of emergence', *Philosophy of the Social Sciences*, 34, 260–82; for a review, see Frantz, T.L. and Carley, K.M. (2009) 'Agent-based modeling within a dynamic network', in Guastello, S.J., Koopmans, M. and Pincus, D. (eds) *Chaos and Complexity in Psychology*, Cambridge: Cambridge University Press.
18. For example, Nowak, A., Vallacher, R., Tessa, A. and Borkowski, W. (2000) 'Society of self: The emergence of collective properties in self-structure', *Psychological Review*, 107, 39–61; Sawyer, R.K. (2005) *Social Emergence: Societies as Complex Systems*, Cambridge: Cambridge University Press.
19. Crutchfield, J.P. and Görnerup, O. (2004) 'Objects that make objects: The population dynamics of structural complexity', *Santa Fe Institute* Working Paper 04-06-020. p. 1.
20. Wright, R. (2000) *Nonzero: The Logic of Human Destiny*, New York: Pantheon Books.
21. For discussion see Richerson, P.J. and Boyd, R. (2005) *Not by Genes Alone: How Culture Transformed Human Evolution*, Chicago: University of Chicago Press; Tomasello, M. (1999) *The Cultural Origins of Human Cognition*, Cambridge, Mass.: Harvard University Press.
22. Vygotsky, L.S., (1981) 'The genesis of higher mental functions', in J.V. Wertsch (ed.) *The Concept of Activity in Soviet Psychology*, New York: Sharpe; Vygotsky, L.S. and Luria, A.R. (1993) *Studies on the History of Behaviour* (edited and translated by V.I. Golod and J.E. Knox), Hove: Erlbaum.

23. Geertz, C. (1962) 'The growth of culture and the evolution of mind', in J.M. Scher (ed.) *Theories of Mind*, New York: Free Press.
24. Vygotsky, *op cit* p. 160.
25. Vygotsky, L.S. (1978) 'Mind in society', in Cole, M., John-Steiner, V., Scribner, S. and Souberman, E. (eds), Cambridge, MA: Harvard University Press. p. 57.
26. Han, S. and Northoff, G. (2008) 'Culture sensitive neural substrates of human cognition: A transcultural neuroimaging approach', *Nature Reviews: Neuroscience*, 9, 646–53.
27. van Geert, P. (1998) 'A dynamic systems model of basic developmental mechanisms: Piaget, Vygotsky, and beyond', *Psychological Review*, 105, 634–77; Doll, W., Fleener, M.J., Trueit, D. and St. Julien, J. (eds) (2005) *Chaos, Complexity, Curriculum, and Culture*, New York: Peter Lang.
28. See contributions in Fogel, A., King, B.J. and Shanker, S.G. (2008) *Human Development in the Twenty-first Century*, Cambridge: Cambridge University Press; also Thelen, E. and Smith L.B. 'Dynamic systems theories', and Fischer, K.W. and Bidell, T.R. 'Dynamic development in action and thought' both in Damon, D. and Lerner, R.M. (2006) *Handbook of Child Psychology*, London: Wiley.
29. Vallacher, R.R. and Novak, A. (2009) 'The dynamics of human experience: Fundamentals of dynamical social psychology', in Guastello et al. (eds) *op cit*.
30. Carruthers, P. (2002) 'The cognitive functions of language', *Behavioral and Brain Sciences*, 25, 657–726.
31. See, for example, Henrich, J., Boyd, R. and Richerson, P.J. (2008) 'Five misunderstandings about cultural evolution', *Human Nature*, 19, 119–37.
32. Mesoudi, A., Whiten, A. and Laland, K.J. (2006) 'Towards a unified science of cultural evolution', *Behavioral and Brain Sciences*, 29, 329–83.
33. Henrich, J. and Henrich, N. (2006) 'Culture, evolution and the puzzle of human cooperation', *Cognitive Systems Research*, 7, 220–45.
34. Boyd, R. and Richerson, P.J. (2005) *The Origin and Evolution of Culture*, Oxford: Oxford University Press. p. 11.
35. Vygotsky, L.S. and Luria, A.R. (1993) *Studies on the history of behaviour*, Hove: Erlbaum (edited and translated by V.I. Golod and J.E. Knox).
36. Scribner, S. (1997) 'Mind in action: A functional approach to thinking', in Tobach, E., Martin, L.M.W., Falmagne, R.J., Scribner, A.S. and Parlee, M.B. (eds) *Mind and Social practice: Selected Writings of Sylvia Scribner*, Cambridge: Cambridge University Press.
37. Johnson-Laird, P.N. (1983) *Mental Models*, Cambridge, Mass.: Harvard University Press.
38. Allman, J. (2000) *Evolving Brains*, New York: W.H. Freeman.
39. Allman, J., Hakeem, A. and Watson, K. (2002) 'Two phylogenetic specializations in the human brain', *The Neuroscientist*, 8, 335–47. p. 335.
40. Gärdenfors, P. (2008) 'The role of intersubjectivity in animal and human cooperation', *Biological Theory: Integrating Development, Evolution and Cognition*, 3, 51–62.
41. Fehr, E. and Rockenbach, B. (2004) 'Human altruism: Economic, neural, and evolutionary perspectives', 14, 1–7.
42. Neiworth, J.J. (2009) 'Thinking about me: How social awareness evolved', *Current Directions in Psychological Science*, 18, 143–7.

43. For a good overview, see Pigliucci, M. (2007) 'Primates, philosophers and the biological basis of morality' (A review of *Primates and Philosophers* by Frans De Waal, Princeton University Press, 2006), *Biology and Philosophy*, 22, 611–18.
44. Pinker, S. (1994) *The Language Instinct*, New York: Morrow.
45. Fitch, R.H., Miller, S. and Tallal, P. (1997) 'Neurobiology of speech perception', *Annual Review of Neuroscience*, 20, 331–53.
46. Piaget, J. (1991) 'Science of education and the psychology of the child', in Light, P., Sheldon, S. and Woodhead, M. (eds) *Learning to Think*, London: Routledge. p. 9.
47. Vygotsky and Luria op cit, p. 111.
48. Vygotsky, L.S. (1962) *Thought and Language* (transl. E. Hanfmann and G. Vakar), Cambridge, Mass.: MIT Press.
49. Epstein, J.M. (2008) 'Nonlinear dynamics, mathematical biology, and social science: Lecture notes', in *Santa Fe Institute Studies in the Sciences of Complexity IV*, New York: Addison-Wesley.
50. Vallacher and Novak op cit p. 374.
51. Richards, D. (1997) 'From individuals to groups: the aggregation of votes and chaotic dynamics', in Kiel, L.D., and Elliott, E. (eds) *Chaos Theory in the Social Sciences*, Michigan: University of Michigan Press.
52. Grüne-Yanoff, T. (2007) 'The explanatory potential of artificial societies', *Synthese*, 169, 539–55.
53. Vallacher and Novak op cit p. 395.
54. Sawyer, R.K. (2005) *Social Emergence: Societies as Complex Systems*, Cambridge: Cambridge University Press. p. 25.

Index

Abouzeid, A., 67
acoustic reception, 101
action concepts, 132
adaptations, 8–10
   as equilibrium (homeostatic) functions, 12
Adolphs, R., 166
Albright, A., 97, 98
Allman, J., 193
Andolina, I., 104
Archaea, 45
arms races, 3
artificial intelligence, 140–1
artificial neural networks (ANNs), 116
   in attractor landscapes, 131, 132, 143
attractors (basins of attraction), 28–30
   in brain activity, 118
   in cognitive systems, 101, 124–8, 134, 150–154, 181
   in complex intelligence, 143
   in epicognition, 181
   in gene regulation, 64
   in physiology, 66
   in vision, 101
attractor landscapes, 101, 132, 134, 151
   brain activity over, 118
   coalitions of, 150–2, 154, 164, 173, 181
   in cognitive functions, 143, 146
   creativity in, 158
   developing, 129, 185, 194
   in knowledge function, 157
   multimodal, 129
audio-visual illusion (McGurk effect), 130
auditory processing, 106–7
autopoietic (theory), 15, 32

Baldwin, J., 84
Ball, P., 45
Barlow, H., 92, 95
Barsch, R., 140
Bartlett, F., 157
Batten, D., 44
Beggs, J., 118
Bénard cells, 32
Blackmore, S. 106
Blakemore, C. 80
Boyd, R., 8, 6, 16, 199–202
brain evolution and environment, 90
brain functions, 102
   in cognition, 143–6
   dynamics of, 110–11
   metaphors of, 90
Bromley, S.K., 59
Bruner, J., 133
Bshary, R., 172
Byrne, R., 174

canalisation, 76–8
   in brain, 80–1, 87–9
   and 'innate' characters, 78
Carruthers, P., 186
cell membrane formation, 57, 60
cell receptors, 65
cerebral cortex, 105
   and brain evolution, 109
   coordinated coding in, 105
   and informational structure, 107
   processing in, 105–8, 130
chaos (chaotic states), 28–34, 117, 196, 143
   in brain states, 118
   in social life, 164, 182, 196
characters (characteristics; traits), 3, 8, 10
   in complex environments, 36, 84
Chase, I., 166
Chemokines, 59–60
Chemotaxis, 46, 161
Chen, L., 114
Chomsky, N., 22, 137, 139, 194
Chyba, C., 37

230 *Index*

circadian rhythms, 51–4
Clayton, N., 167
Cleland, C., 37
cognitive functions 6, 112–13
 ANN models of, 115–16, 119
 complex, 10
 deep structure and, 144
 dynamic theory of, 117–20
 emergence of (from neural activity), 124
 evolution of, 113
 'extended', 153
 fragmented theory of, 114, 135–8
 'higher', 143
 as intelligent systems, 133
 in invertebrates, 113–14, 136
 new views of, 152–4
 in relation to cultural tools, 185
 social, 26
 in thinking, 140–1
combinatorial regulation, 75
complexity, 2
 definition of, 19–20
 describing, 22–25, 30–1, 34
 dynamic, 25–8
 environmental, 21, 51, 109–13
 as 'heterogeneity', 12
 increase in evolution, 2–4, 62
 in intelligent systems, 4
 and linear models, 27–8
 neglect of, 11, 13–15
 origins of (in living things), 31–5
concept formation, 125–8
 from neural activity, 121–3, 134
construction of multisensory images, 129–30
construction of visual images, 121–3
Cosmides, L., 9, 11
Craft, W., 95
creationism (intelligent design), 5, 20
Crespi, E., 79
Crick, F., 38
Criticality, 30, 34, 63, 117–18, 143
 in cognitive functions, 117
Crutchfield, J., 34, 182
cultural tools, 180–3
 in child development, 185
 in cognitive differences, 185
 in differential brain activity, 185
 and enhanced cognition, 186, 188–200
 and metacognition, 186
 as psychological tools, 184
 and reflective abstraction, 185
culture in humans, 182–5
 as basis of human creativity, 195
 brain specialisations for, 192
 and neo-Darwinian biology, 186–8
 chaotic activity in, 195
culture, in primates, 169

Dale, R., 117
Darwin, C., 1–6, 9,10, 36, 39, 76
 and psychology, 5, 9
Dawkins, R., 3–4, 187
DeAngelis, G., 104
Delauney, F., 54
Dent-Read, C., 14, 77
Denver, R., 79, 82
Depew, D., 17
determinism, 14, 16, 26–8, 33
development, 61–75
 assembly and growth model of, 87
 of body axis, 92
 of brain, 90–1
 canalised, 76–9
 and ecology, 83
 epigenetics of, 93
 evolution and (evo-devo), 88
 gene recruitment in, 91–3
 as lifelong process, 84–8
 teleology in, 87
developmental plasticity, 78–83
 in brain, 80–1, 87–9, 109–10, 152
 transgenerational effects of, 82
differentiation (in development), 58–9, 71–6, 162–6
DNA, 8, 38–42
Dunbar, R., 171
dynamic systems approach, 16
 in brain, 110
 in cognition, 150
 in social cooperation, 181, 185
 theory (DST), 17, 29–33
 in vision, 110

Edleman, G., 15
Elman, D., 139
emergence, 30–4, 66, 83, 118, 129, 147, 153, 182
Emery, N., 167
emotional intelligence, 192
environmental change, 10–12
   and brain evolution, 90
   and complex intelligence,11
   and complexity, 12–14
   c.w. 'Darwinian' environments, 10
   and heterogeneity, 12
   neglect of (in theories), 14
   structure in, 11, 21
   tendency to simplify, 14, 18
environments, describing, 3–4, 12–15, 26–7
   stable (in EP theory), 10
epicognition, 171–2, 180–1, 182
epidermal growth factor (EGF), 59
epigenetics, 75–8, 82, 83
event concepts, 131
evo-devo, 17, 71
evolutionary mechanisms, neglect of, 7
evolutionary psychology (EP), 9–10
extracellular matrix, 59
Eysenck, M., 140, 141

feelings/emotions, 145
   culturally modulated, 192
   integration with cognition, 146
Feldman, J., 181
Fodor, J., 6, 113, 135, 141
Fox, S., 44
fractal structure, 31, 106, 118, 153
Freeman, W.J., 111, 117, 120, 146

Gaia (theory), 16
Gardner, H., 77
Geiger, B., 61
gene numbers, in evolution, 62, 83
genes, 8, 14, 20–1, 60–3
   alternative splicing from, 62
   in altruism, 194
   in cognitive functions, 152
   as designers, 38
   determinism versus dynamics in, 62, 78
   in development, 70–4

   evolutionary significance of, 159
   in interactive networks, 62–3
   in language, 194
   new concepts of, 62–3
   in origins of life, 39–40
   as resource for metabolism and development, 4
   in self-organising dynamics, 63
   transcription, 77
   utilisation of, 61–2
Gestalt theory, 105, 140
Giurfa, M., 136
Godfrey-Smith, P., 12
Goldberger, A., 67
g-protein-coupled-receptor (GPCR), 59
Gregory, R., 128
Grüne-Yanoff, T., 196
Guerin, S., 165

Haldane, J.B.S., 39
Harnad, S., 135
Hebb, D., 137
Hoffman, D., 93
holistic (models), 15, 17
human evolution, 176–9
   brain size in, 178–80
   cultural tools in, 179, 180–2, 183
   pre-adaptations for, 177
human language, 194–5
   and human evolution, 195
human personality, 194
Hurri, J., 100
Hypothalamus, 66–8, 144
Hyvarinen, A., 100

imitation learning, 168
information, 24–5
   and complexity, 34–5
   parameters of, 25, 65, 101–2, 108
   spatiotemporal, 31
   and structure, 13, 19, 25
   as structural parameters, 113–14, 121–4, 126–7, 155–6
intelligence, in early evolution, 43
   in cell motion, 57–62
   in cell signaling, 56–67
   c.w. cue-response mechanism, 37, 44–7, 51, 55, 114
   in single cells, 45–55

intelligent design, 5
inter-emotion, 193
  and altruism, 193
  compared with sociobiological view, 193
inter-subjectivity, 193

Jablokna, E., 18
Jackson, S., 132
James, W., 93

Kaneko, K., 151
Kauffman, S., 40
Keane, M., 140
Kerzel, D., 131
knowledge, nature of 7, 139–40
  as cultural tool, 188–9
  dynamic view of, 157
Kuffler, S., 92
Kunkle, D., 165

Lamarckian inheritance, 82
Lamb, M., 18
Lancet, D., 40
Langton, C., 44
Lappin, J., 95
lateral geniculate nucleus (LGN), 103–4
Laudet, V., 54
learning, 6, 136–9, 151
  in bacteria, 50–2
  in cells, 45–6
  as dynamic system, 154–6
  and environmental complexity, 109
  in invertebrates, 114
  learning theory, 137–8
  in neural networks, 116
  in sensory systems, 129
Lee, T-S., 106
Lei, H., 102
lifelong development, 85–9
  in behaviour and cognition, 85
  in brain, 86, 109–10
ligands, 47
Lorenz, E.N., 30, 31
Love, A., 14, 15
Lovelock, J., 15–16
Luria, A.S., 195

Mackay, D., 105, 108–9
Martin, P., 91, 92
Masland, R., 91, 92
Mateos, R., 27
Maturana, H., 15–16
Mauron, A., 20
McGurk, H., 130
meaning of meaning, 128
mechanical models (of living systems), 14, 6, 26, 142, 152
memory, 138–9
  as cultural tool, 190–1
  dynamic view of, 155–7
messenger RNA (mRNA), 61
  in development, 75
  In gene transcription, 61
  regulatory functions of, 61–2
metabolisms, primordial, 40
Miconi, T., 3
Miller, S., 39
mirror neurons, 132, 170
Mitchell, M., 7, 19
modules ('mental organs'), 9–10
  in metabolic networks, 58, 150
Morita, K., 107
Müller, G., 70
multi-agent systems, 181–2
music, 7
  lack of explanation of, 7
  psychology of, 107, 132
  structure in, 22, 28, 183

Nader, K., 138
Nakagaki, T., 45
natural selection, 8
  acceptance of, 1–2
  applied to cognitive functions, 9–10
  cell dynamics and, 64
  complex environments and, 11
  integration in, 13
  in origins of life, 42–4
  'purpose' in, 13
neo-Darwinian theory, 4
networks, 25
  coalitions of, 144
  in cognition, 112, 121, 127–129, 137, 143, 146
  developmental, 83–4

networks – *continued*
  in evolution, 152
  gene interaction, 61–2
  metabolic, 26–40
  in nervous systems, 86–7, 101, 103, 106–10
  reflective abstraction in, 150–2
  signalling, 57–63
  social, 170–5, 185–7
neuroendocrine stress axis, 65–6, 79
  transgenerational effects of, 80
neurons, 69, 80, 88–90, 97, 110, 112–13
  computations in, 107
  connectivity of, 152, 175
  model, 115
  multisensory, 130, 132
  in relation to cognition, 119, 122
Noë, A., 91, 95, 133
noncoding RNA, 62
Nowak, A., 197

O'Regan, K., 95, 133
olfactory processing, 120
olfactory reception, 101
Olveczky, B., 100
Oparin, A., 39
origins of life, 37–8
  autocatalytic sets and, 40–1
  before genes, 38–9
Oyama, S., 13, 14, 20

Paz-y-Miño, C.G., 168
Peelen, M., 128
percepts, dynamic creation of, 124–8
Pessoa, L., 147
Physiology, 64–7
  homeostatic model of, 64
  nonlinear dynamic models of, 66–7
Piaget, J., 147–50, 195
  and reflective abstraction, 150
Pigliucci, M., 14, 70
Pincus, D., 152
Pinker, S., 7, 9, 10, 115, 139, 194
Poincaré, H., 27
point light stimuli, 94–5, 125–7
Potts, R., 21
Premack, D., 175
Premack, A.J., 175

Prigogine, I., 32
problem-solving, 141, 142

reasoning abilities, 141
reasoning (IQ) tests, 141
reductionism, 14–15, 17, 34
reflective abstraction, 150
  in motor development, 151
  in multi-agent systems, 182
  in networks, 150–1
  in non-human animals, 150
  in social context, 185–6
reproduction, 41
  asexual, 44, 71
  in origins of life, 41–2
  sexual, 72
retina, 96–101
retinal ganglion cells, 96, 103
Richards, D., 197
Richerson, P., 6, 12, 21
Rock, I., 93

Sabelli, H., 67
Sawyer, K., 182, 197
science, as cultural tool, 189–90
Scribner, S., 191–2
self-organised systems, 30–4, 72, 83, 87, 117, 118, 126, 187
sensorimotor transformation problem, 132
serial ligation, 48
Shenhav, B., 40, 41
Shepard, R., 140
Shiffrin, R., 6, 138–9
Shvartsman, S., 59
signaling systems, 49, 54, 57–61
  in brain networks, 87–8, 104, 109, 112, 179
  in cognition, 150
  in development, 72–82
  as dynamic regulations, 63, 110, 112
  in gene use, 61–3
  in physiology, 64–7
  social, 161–2
Singleton, K.L., 60
Sinha, C., 6
Smith, J.M., 38
social aggregation, 160
  in bacteria, 161–2

social brain, 170–1, 192
social complexity, 166, 167
　and cognitive functions, 168, 176, 179
　and epicognition, 171–2, 179–81, 182
social cooperation, 160
　brain specialisations for, 192
　cognitive demands of, 163, 164, 166
　dynamics of, 165, 166
　in fish, 166
　in humans, 176
　in hunting, 172–3
　in social insects, 163–4
Spitzer, N.C., 69
Spivey, M., 117
Stanley, G., 104
Stearns, C., 78
Stein, L.A., 153
Stephen, D., 118
Stevens, C., 73
Stewart, I., 26, 33
Stoner, G., 97, 98
structure (environmental), 9, 11, 13
　and complexity, 19
　and correlation (association), 23
　deep (complex), 30–1
　describing, 21–3
　emergent, 33–5
　and intelligent systems, 34–5
　and predictability, 23
　spatiotemporal (dynamic), 23–4
Sur, M., 80
symbol grounding, 128–9
Szathmary, E., 38

tactile reception, 101
Thelen, E., 151
thinking, 139–40
　'logical', 141
　computational model of, 140
　as cultural tool, 191–2
　dynamic view of, 158–9
　'theory of mind', 167–8
Tooby, J., 9, 10
Turing, A., 32, 71, 114

Vallacher, R., 197
Van Sluyters, R., 80
vision, 91
　dynamics of, 101, 104
　and environmental structure, 93–6
　feature detector (theory), 92, 97, 99
　and informational structure, 100, 105–6
　the 'snapshot' view of, 91
visual illusions, 128
von Helmholtz, H., 98
Vygotsky, L., 183–5, 195

Waddington, C.H., 76
Wallach, H., 94
Watson, J.D., 38
Werblin, W., 99–100
Wolpert, L., 72
Wuensche, A., 64
Wynne, C., 150

Zukow-Goldring, P., 14, 77, 181